"In evocative and moving prose, *The Shifting Sands of the North Sea Lowlands* takes us down rivers and along coasts, through mudflats, silt, and fenlands, to explore the remarkable littoral landscapes and literary imaginations of the North Sea Lowlands. In Ritson's reading, these ever changing places and the diverse texts they have animated and inspired offer up an incredible richness of possibilities both for understanding our pasts and for crafting shared futures in this Anthropocene epoch."

Thom van Dooren, Department of Gender and Cultural Studies and Sydney Environment Institute, University of Sydney, Australia

"Where does the Anthropocene tell its stories? And who are the storytellers? With the skill of a novelist and cutting-edge trans-disciplinary methodology, Katie Ritson finds the answers to these questions in the eloquent landscapes of the North Sea Lowlands, where sands and waters, nature and history, human creativity and ecological predicaments merge into one another. Vis-à-vis with the combined imagination of elements and literature, *The Shifting Sands of the North Sea Lowlands* is one of the best examples of how the environmental humanities can contribute to shaping the generous imaginaries we need in a time of shifting horizons."

Serenella Iovino, co-editor of Environmental Humanities: Voices from the Anthropocene *and* Italy and the Environmental Humanities: Landscapes, Natures, Ecology

The Shifting Sands of the North Sea Lowlands

Global seawater levels are rising and the low-lying coasts of the North Sea basin are amongst the most vulnerable in Europe. In our current moment of environmental crisis, the North Sea coasts are literary arenas in which the challenges and concerns of the Anthropocene are being played out.

This book shows how the fragile landscapes around the North Sea have served as bellwethers for environmental concern both now and in the recent past. It looks at literary sources drawing on the countries around the North Sea (Denmark, Germany, the Netherlands, and England) from the mid-nineteenth century onwards, taking them out of their established national and cultural contexts and reframing them in the light of human concern with fast-changing and hazardous environments. The six chapters serve as literary case studies that highlight memories of flood disaster and recovery, attempts to engineer the landscape into submission, perceptions of the landscape as both local and global, and the imagination of the future of our planet. This approach, which combines environmental history and ecocriticism, shows the importance of cultural artefacts in understandings of, and responses to, environmental change, and advocates for the importance of literary studies in the environmental humanities.

This book will be of great interest to students and scholars of the Environmental Humanities, including Eco-criticism and Environmental History, as well as anyone studying literature from the Germanic philologies.

Katie Ritson is Managing Editor at the Rachel Carson Center for Environment and Society, LMU Munich.

Routledge Environmental Humanities

Series editors: Paul Warde (University of Cambridge, UK) and Libby Robin (Australian National University)

The *Routledge Environmental Humanities* series is an original and inspiring venture recognising that today's world agricultural and water crises, ocean pollution and resource depletion, global warming from greenhouse gases, urban sprawl, overpopulation, food insecurity, and environmental justice are all crises of culture.

The reality of understanding and finding adaptive solutions to our present and future environmental challenges has shifted the epicentre of environmental studies away from an exclusively scientific and technological framework to one that depends on the human-focused disciplines and ideas of the humanities and allied social sciences.

We thus welcome book proposals from all humanities and social sciences disciplines for an inclusive and interdisciplinary series. We favour manuscripts aimed at an international readership and written in a lively and accessible style. The readership comprises scholars and students from the humanities and social sciences and thoughtful readers concerned about the human dimensions of environmental change.

The Shifting Sands of the North Sea Lowlands

Literary and Historical Imaginaries

Katie Ritson

Routledge
Taylor & Francis Group
LONDON AND NEW YORK

earthscan
from Routledge

First published 2019
by Routledge
2 Park Square, Milton Park, Abingdon, Oxon OX14 4RN

and by Routledge
52 Vanderbilt Avenue, New York, NY 10017

Routledge is an imprint of the Taylor & Francis Group, an informa business

Publisher's note: This book is a revised version of the dissertation submitted by the author at Ludwig-Maximilians-Universität Munich in 2016.

British Library Cataloguing-in-Publication Data
A catalogue record for this book is available from the British Library

Library of Congress Cataloging-in-Publication Data
A catalog record has been requested for this book

ISBN: 978-1-138-59110-3 (hbk)
ISBN: 978-0-367-07721-1 (pbk)
ISBN: 978-0-429-49064-4 (ebk)

Typeset in Goudy
by Wearset Ltd, Boldon, Tyne and Wear

Contents

Figures

Acknowledgements

This book has been made possible in no small part by the Rachel Carson Center for Environment and Society at the Ludwig-Maximilians-Universität (LMU) in Munich. The RCC is a hub for scholars of all shades who are interested in the environment; it is simply not feasible to list here the names of all the fellows and associates whose energy, enthusiasm, and wisdom have played their part in inspiring this project and its author. It is a wonderful place to work and I look forward to more inspiring encounters in the future. The Institute for Nordic Philology at LMU in Munich gave me my second institutional home, and has provided steady encouragement and camaraderie along the way.

Especial gratitude is due to Eva Jakobsson, Dolly Jørgensen, and Finn Arne Jørgensen, who between them arranged two separate stays at the University of Stavanger in 2014 and 2018. Without their help this book would be very different. Thank you to Øyvind Rimbereid and Kjersti Vik, who made time to meet and discuss their work with me in person.

A few other people do need to be singled out for their particular help and support with this endeavour. Some of them took time to give me advice on research or writing, or pass on tips for combining scholarship with the demands of family life. Sometimes just the fact of their taking this project seriously was more help than they ever realized. Thank you Melanie Arndt, Greg Bankoff, Anthony Carrigan (RIP), Hanna Eglinger, Rob Emmett, Frederike Felcht, Axel Goodbody, Arielle Helmick, Isabel Hoving, Benedikt Jager, Matt Kelly, Claudia Leal, Bernhard Malkmus, Diana Mincyte, Ruth Morgan, Ursula Münster, Roger Norum, Chris Pastore, Ruth Oldenziel, Karen Oslund, Cindy Ott, Libby Robin, Lise Sedrez, Laura Sayre, Nora Simonhjell, and Sabine Wilke. Special thanks go to the wonderful RCC editors past and present—Brenda Black, Steph Hood, Hannah Roberson, Sam Rothbart, and Harriet Windley—for their unstinting goodwill and attention, and to Paul Martin for his meticulous copyediting.

My greatest debt of gratitude is due to my two mentors, Annegret Heitmann and Christof Mauch. Their intellectual inspiration and their generous encouragement are what kept this project going through all its various ups and downs. Thank you both.

And last but not least; thank you Tobi, for your patience; and Georg, Samuel, and Anton, for your impatience. The terrain has not been easy—treacherous tides, silt, shifting sands … I'm so glad you came with me!

Introduction

On the edge of the North Sea

I am on the top deck of a double-decker bus, waiting and watching for the moment when Ely Cathedral first looms up from the Cambridgeshire Fens, an indistinct shape of dark grey on grey, difficult to discern on the horizon. The bus lurches along narrow country lanes, its upper windows lashed by tree branches, brushing against hedgerows that open out to reveal vistas of flat land cut into parcels by ditches. Whatever the weather outside, from my seat on the top deck, the effect is always the same: a faded landscape of grey-green, blurred by the dirty glass and the clumsy motion of the bus.

When I was an undergraduate in Cambridge and felt the need to escape from the cloistered world of my college, I would often catch the bus to Ely. My room looked onto the noisy Drummer Street bus station, and so I knew the times of the buses that left for Ely once an hour during the day; the £4.20 fare to Ely was the price of an afternoon of what felt like freedom, a freedom that started as soon as I had left the busy streets and turrets of the city behind. I would stare through the window for the whole of the journey, enjoying the sense of being the one looking. It was a relief to be unobserved after the fishbowl of college life.

The grey bulk of Ely Cathedral—the "Ship of the Fens," as it is often called—really does call to mind a ship on the horizon, especially when viewed from a moving vehicle through a sheen of grime. As soon as I caught that first glimpse of it, I could drift away into a pleasurable imagination of landscapes past, when Ely was an island named for the eel-rich waters that surrounded it, at a time when humans understood the wealth of the muddy and mutable enough to build their cathedrals in such places. The Fens were wild wetlands once, and the human inhabitants moved around by boat and on stilts, effectively living half in the water. Sometimes my imagination would lead me in the other direction, and I pictured the inundation that would eventually return to Ely, waters that would swirl and carry away the peaty soils; the cathedral would begin to list, like a ship in the swell, its walls to crumble, and the waters would rise higher. Eventually just a suggestion of its presence—an arch, a broken spire—would be visible above the churning grey waves.

At that time, in the early noughties, when the spectre of climate change was inserting itself ever more forcefully into the optimism of our privileged

Figure I.1 Towards the "Ship of the Fens."
Source: photo by John Sutton. [cc-by-sa/2.0 – © John Sutton – geograph.org.uk/p/5760982].

twenty-something lives, there was something perversely comforting about this reverie I regularly sought out. Ely Cathedral, monstrous and beautiful, was constructed in a landscape that had always been in a tense dialogue with the North Sea. Humans had reclaimed and drained and built, while the sea had eroded and flooded and permeated. Other spires had already disappeared beneath the waves and some day, I supposed, new ones might rise up and fall again. The great uncertainty around what we humans had unleashed in our exploitation of planetary resources found its match in the sure knowledge that, ultimately, the agency of the natural world would outlast our own; and that there were places where this fact was distilled and visible. Ely was just such a place.

More than ten years after I left Cambridge, I embarked on a return journey to these memories and began to explore what the image of the listing stone ship in the landscape might mean, not just to me, but to others too. The North Sea lowlands are rich in sites of memory where the fusion of natural and human agencies across multiple timescales have made a powerful imprint on the cultural imagination. Iconic assemblages such as Dunwich, Rungholt, and Denmark's "buried church," to name but three, are part of the sediments of our shared human imagination, articulated in physical but also in literary form. In this book I have attempted to investigate how the North Sea lowlands serve as a site of human imagination, and to consider how our present historical moment might make these watery places into particularly rich landscapes for us to think with, and about, both the past and the future.

The lowlands surrounding the North Sea basin, extending from England's Wash in the northeast to the Danish coast of Jutland in the northwest and down to the polders of the Netherlands in the south, can just as easily be described as waterscapes as they can landscapes. The low-lying regions around the shallow basin share a vulnerability to tidal movements and storms and a common history of human engineering. On the Frisian coasts and in the English estuaries of the Wash, beyond the sandy beaches, vast expanses of mudflats open out at low tide. Neither completely sea nor land, these liminal, shifting spaces abound in sustenance for vast populations of wetland birds, but have been historically treacherous for humans. Behind the dykes, in the drained lands of the Netherlands, Belgium, and the East Anglian Fens, the ground is flat and saturated, divided into green and angular parcels by the ditches and sluices that criss-cross the marshy ground. Today, this ground is shrinking, the water looking for ways to return. On the coasts, the border between the dry land and the sea is constantly shifting, not just with the daily tides, but with the gradual erosion of coastlines and the construction of new barriers.

Humans have lived in the North Sea lowlands for millennia. Natural and human forces have constantly redrawn the map of this tidal basin, as the physical forces that govern the tides and the slow formation and movement of silt and sandbanks have been met by human ingenuity in responding to the risks of a dynamic environment. For many centuries, during which there were few roads and no motor vehicles, seas and rivers were the sole routes of communication and transport between regions and countries, and the North Sea basin's waterways became a central thoroughfare in northern Europe. Boats of all kinds traversed it in all seasons, and knowledge of the coasts and the tides that could carry craft far into inland ports was a feature held in common by the cultures of these regions. Coastal settlements and river towns were lively places where messages and goods were exchanged, and many people lived prosperously on the profits from trade, fishing, and often enough piracy and the looting of shipwrecks. In the nineteenth and early twentieth centuries, the coasts boomed as people took to the new notion of the seaside resort, and better-off citizens of the inland flocked to the littorals to enjoy the sea air and bathing (Pye 2015; Pott 2003; Parker 1978; Morley 1968).

It is only comparatively recently that industrialized societies have turned from dependence on such waterways to a new dependence on land and air routes, whilst simultaneously elevating the nation-state to the main unit of political and economic development. The North Sea coasts have been recast as peripheries, areas set quaintly on the margins of national territories, no longer the central hubs of economy and encounter they were just a few hundred years ago. Small and large ports have closed down, coastal towns have declined. The discovery of oil under the North Sea has brought prosperity to those national economies with a stake in the seabed, but fossil fuel technologies have become

Figure I.2 Map of the North Sea basin.

Source: courtesy of Tobias Schiefer.

embedded into globalized market forces at the expense of local ones, and sea and land are suffering the effects of pollution, both regulated and otherwise. Offshore and onshore windfarms are a visible sign of newer technologies that utilize the lowlands as an energy resource. The process of land reclamation, in the past driven by the desire for more arable and grazing land for agriculture, has almost entirely ceased around the North Sea, with even the attempt to maintain existing lowlands beset by difficulties. Seawater levels are rising, and the current lines of the coast, more low-lying than ever due to land shrinkage, are being reclaimed by the water, while human settlements are forced to move further inland (Gerdes 2011; Fernandez-Bilbao 2012). The unstable and diminishing frontiers between land and sea now showcase the ecological dynamism that humans previously tried to subdue: there are Wadden Sea National Parks in Germany, the Netherlands, and Denmark, and the Wadden Sea has also been declared a UNESCO World Heritage Site, while there are myriad smaller wetland reserves around the North Sea coasts in all bordering countries.

The lowlands are a quintessential example of a landscape in a process of constant flux, contested by the various natural and cultural forces that play out around the edges of the North Sea. The littorals form a cultural space with a shared history of flooding and risk management and many traditions in common. The environmental historian Greg Bankoff has argued for the entire North Sea lowlands to be seen as a shared risk culture, incorporating not just the Frisian coasts of Germany and the Netherlands, but also the coasts and fens of eastern England and the southwest coast of Denmark:

> Indeed, the Netherlands, the northwest coast of Germany and the western coast of Denmark form together with eastern England one vast North Sea Basin system continually shaped and reshaped by the processes of storm, flood and erosion. Over time, the people living along its shores have had to find ways of accommodating to this dynamic world by adapting to its rhythms, learning to live with its risks, taming the worst of its excesses and exploiting its resources for their own wellbeing and prosperity.
>
> (Bankoff 2013, 4)

A literary exploration of this landscape focuses on these cultural aspects of the lowlands. My preference for the term "landscape" is deliberate. Landscape (or indeed waterscape) is, as Simon Schama showed us in his influential *Landscape and Memory*, always a cultural term (1995, 10); in the introduction to *The Conquest of Water*, David Blackbourn notes that "landscapes are both real and imagined" (2007, 18). Cultural geographers such as Daniel Weston have called for an approach to landscape that treats it, not as something static, but as a dynamic, discursive process between subject and object, writing that landscape "signifies both the material place and its pictorial representation" (Weston 2011, 175), and that literary texts, along with other art forms, are capable of exploring the tension between the objective place and the subjectivity of its framing. Literary texts engage with the human experience of space. They show

how space is invested with meaning, employing the imagination in much the same way as the word "landscape" does—inviting us to admire a view that is, to a lesser or greater extent, a reflection on humanity's place in the world. The "scape" of landscape—drawing on Arjun Appadurai's idea of scapes as the building blocks of "multiple worlds which are constituted by the historically situated imaginations of persons and groups spread around the world" (Appadurai 1990, 296)—indicates its abstractability and transferability from the physical world; literary texts make an important contribution in this process of mediation between physical "land" and its imagination. The nexus of real and imagined place in the word landscape is central to the discussions of environmental historians and others working across the environmental humanities. Literary works do not just capture a particular landscape for posterity, they reflect on it discursively. This literary reflection can take many forms: it can look critically at current or historical interactions with space, suggest ways of interacting with landscapes and future generations, invest a non-human space with human emotions—in short, literary works explore the complex network of factors that comprise the human understanding of space and place.

The rise of environmental history as a distinct scholarly field has shown the importance of historical analysis that looks across the borders of nation-states to the underlying environmental factors that shape the development of areas and regions. This topographically oriented approach, however, is not a typical feature of literary studies, even if postmodernism's spatial turn has led to a broader interest in the category of literary space. The aesthetic dimension of the North Sea lowlands—the cultural representations of its liminal nature, caught between land and sea—has been studied most frequently in the context of local or historical studies of particular artists or writers (for example in studies on the writer Theodor Storm and the poet John Clare).

As a comparativist, I believe in the value of looking at literary texts across national and linguistic borders, even when these texts are considered parts of differing literary and cultural traditions. As the geographer John Agnew has written, "the taken-for-granted nationalism that has long afflicted the practitioners of the emerging social sciences has meant that the nation-state has become regarded as not simply the main but the only geographic unit of account" (Agnew 2011, 323). Similarly, Ulrich Beck has called for transnational and cosmopolitan methodologies as part of a process he calls "reflexive modernization" (Beck 2003, 13). Agnew and Beck were not addressing themselves directly to scholars of literature, but literary studies may be called to task in the same breath; freeing literary texts from their established cultural contexts is an opportunity to see them with fresh eyes. Looking at the aesthetic representation of landscape across different literary traditions has the benefit of foregrounding the environmental factors that are held in common, particularly in our increasingly globalized modern world, and bringing to light discourses of ecological places and processes that are framed and understood through a specific locality. It also forces an engagement with current debates about the values we attach to physical places and how these relate to our sense of the wider world. While technological developments (such as the internet and air travel) mean that we are becoming increasingly independent of our physical location, the spectre of climate change forces us to confront our very tangled relationship with the ecosystems we inhabit.

The literary landscape in the age of humans

The question of why it might be worthwhile to look comparatively at the literary incarnation of the North Sea lowlands is closely tied to another question: "Why now?" For while the landscapes described above are presented primarily in spatial terms, they are also distinctly temporal phenomena. The North Sea coasts have always existed in a state of flux, and the current threat of anthropogenic climate change is only the latest in a series of large-scale reshapings that have taken place over millennia. Archaeologists have shown that, into the last ice age, a shelf of dry land (Doggerland) connected the three sides of the sea basin and was slowly inundated as the ice sheets melted and seawater levels rose (Ward *et al.* 2006; Gaffney *et al.* 2009). Throughout the estimated 7,000 years that have passed since the topmost ridge of Dogger Bank disappeared beneath the waves, the visible coastline has been consistently changing. Over many centuries, human engineering has "reclaimed" arable land from the waves and built prosperous settlements around river ports, while storm surges and silt movement have interrupted or destroyed human habitations. Humans have been drawn to the coasts because of the rich resources on offer, whether this has been the fish stocks (the annual herring season has long been a lucrative source of income on the North Sea coasts), peat, land for grazing or for arable crops (see Pye 2015, 169–91), or beachfronts to be developed into tourist resorts.

The way that these processes of change are depicted in writing can show how human inhabitants relate to this landscape and their shifting place in it. All landscapes change over time, but the North Sea coasts are amongst the geologically youngest and most dynamic landscapes in Europe. The coastlines have changed, sometimes quite dramatically, within recorded memory—even within living memory—whether by human engineering and ingenuity or by natural forces. Historical accounts of storm surges (for example, the two "Grote Mandränken," the "great drownings of men," in 1362 and 1634 in Germany, and similar catastrophic events) have been formative in the regional identity of the lowlands (Rieken 2005). In the Netherlands, for example, the country named for its watery situation, the coastlines of the low-lying islands in the Rhine Delta region were altered by the devastating storm surge of 1953 and the subsequent construction of the Delta Works, designed to prevent large-scale catastrophes in the future (see, for example, Gerritsen 2005; Hall 2013). The history of these landscapes is not just the history of extreme tides, but also of constant low-level flux, in which the slow silting up of rivers and the appearance of sandbanks has been just as important for the waxing and waning of settlements and trade routes as the sudden breaching of the dykes during storms (for a long account of how the silting up of a natural harbour over hundreds of years led to the decline and eventual obliteration of a medieval town, see Rowland Parker's *Men of Dunwich* 1978). The ways in which human inhabitants have adapted to both long- and short-term changes in their environment makes the history of this region of particular interest in our current moment, when strategies for adapting to and mitigating the effects of climate change are being discussed in town halls as much as they are at international governmental summits.

Recent history has seen a shift from large-scale plans for further land reclamation to measures designed primarily for the protection of existing human settlements. The Zuiderzee project was begun in 1919; Adolf Hitler opened the reclaimed Adolf-Hitler-Koog (now known as Dieksanderkoog) in 1935; and, as recently as the 1960s, there were plans to drain the Wash—the vast area of mudflats between Norfolk and Lincolnshire—and develop two new shipping ports there (the latter exciting possibility is discussed in George Morley's 1968 book *The North Sea* in the final chapter, "Prospects for the Future"). However, more recent large-scale projects, such as the Delta Works and the Thames Barrier in London, have a more protective function (Kelly 2018). Latterly, concepts of "managed retreat," ecological conservation, and rewilding (e.g., Barkham 2014; see also Mabey 2005) have influenced the management of the lowlands. Besides the tourists who still come to the North Sea for bathing and beach pursuits, there are ecological tourists in search of rare birdlife and sightings of seals. Konik horses, descendants of the European tarpans, have been (re) introduced to the Norfolk Fens and to Oostvaardersplassen in the Netherlands to keep down vegetation and preserve the wetland ecology, replacing the drive for ever more agricultural land (see e.g., Mabey 2005, 82–90). Similarly, the Wadden Sea coast in Germany and Denmark was given protected conservation status as a UNESCO World Heritage Site in 2009, recognition of both its vulnerability and its cultural significance (Wöbse 2013; Hasse 2011).

Nature protection has become a leading consideration in debates about the region's future. The Netherlands, Germany, Denmark, and the UK are all signatories to the Ramsar Convention on wetland protection, and boast more than 300 designated Ramsar wetlands between them.[1] This politicized attitude to wetlands represents something of a *volte face* since, for much of recorded history, wetlands have been regarded as unsavoury, disease-ridden spaces that could be improved by drainage (Rotherham 2013, esp. 25–47; Blackbourn 2007, 45; Giblett 1996). Carolyn Merchant notes the significance of wetlands as players in conservation debates:

> The marsh—drained for pasture, blamed for disease, and exploited for wildfowl—has today become a cause célèbre of ecology. The diversity and interdependence of its complex food chains, saline balances, and irreplaceable flora and fauna are important elements in the ecologists' argument that the preservation of diversity and genetic variability is essential to the survival of life on the planet.
>
> (1980, 61)

The North Sea lowlands have been the subject of several shifts in terms of perceptions of the resilience of the landscape and the necessity for human efforts to engineer and protect it. This history of flux, of the push-and-pull between human engineering and geological dynamism (and the intellectual movements that have accompanied this flux), becomes more urgent in light of the fact that the North Sea coasts are confronted with a horizon of new changes over the coming centuries. Anthropogenic climate change is accelerating and altering the processes that are permanently underway in this unstable region. Predicting the rate and impact of sea level rises remains difficult (Gerdes 2011), but as seawater levels rise, the rate of coastal erosion will increase and the damage caused by storm surges will have greater reach and frequency. Many communities have already started to respond to this new threat by raising dykes or building new barriers, but, increasingly, strategies of mitigation and (willing or unwilling) retreat are becoming a more typical response (see newspaper reports in the *Flensburger Tagesblatt* [Maas 2013], *Guardian* [Barkham 2015] and the suspiciously positive article in *Die Welt* [Hillmer 2013]).

The North Sea lowlands thus represent a diverse landscape of great vulnerability, both historically and in the future. In our modern-day context of climate change and environmental degradation, they can serve as a barometer for sensitivity to discourses of environmental crisis. The fact that these landscapes are often heavily engineered and adapted to human settlement add to their interest as objects of study. Some areas of the Netherlands lie as much as 6.5 metres below sea level, and humans can only live there at all thanks to the complicated system of constructed dykes and sluices that keep the land dry and desalinated enough for agriculture. The mixture of human ingenuity and adaptation and geological vulnerability allows these landscapes to stand as microcosms of the human predicament in the recently recognized Age of Humans—the Anthropocene.

The Anthropocene is a contentious term, but the discussions by scholars about the wisdom (or otherwise) of adopting it have been eclipsed by its popularity in public debate. The tension that this contentiousness provides—about the extent of human intervention in "natural" processes, the utility of thinking in geological timescales, and the problems inherent in an ecological view of humanity as a single biological species—makes the concept productive in a literary context. Literary texts about the North Sea lowlands contain and reference complex discourses, and articulate many of these difficulties and paradoxes. The dynamic landscape reflects both the extent and the limits of human agency, and highlights the threat to the future that anthropogenic environmental change poses. Thus, the exercise of taking up an Anthropocene perspective is fruitful for the kind of literary analysis in this project.

The term Anthropocene was coined by scientists Eugene F. Stoermer and Paul Crutzen in the 1990s. In their short paper they wrote:

> Considering … major and still growing impacts of human activities on earth and atmosphere, and at all, including global, scales, it seems to us

> more than appropriate to emphasize the central role of mankind in geology and ecology by proposing to use the term "anthropocene" for the current geological epoch. The impacts of current human activities will continue over long periods.
>
> (Crutzen and Stoermer 2000)

The naming of the Anthropocene represents a new rupture in the geological history of the planet and, uniquely, one caused by the activity of *Homo sapiens*. The start date of the Anthropocene remains somewhat open: Stoermer and Crutzen suggest the Industrial Revolution at the end of the eighteenth century as a starting point (17), when the invention of the steam engine triggered a widespread dependence on fossil fuels to meet new energy needs. They allow, though, that other starting dates are also possible, such as the "Great Acceleration" of the 1950s or the invention of agriculture. Will Steffen lists the year 1950 as the main alternative candidate for the starting date (Steffen 2013, 490). The basic idea that the geological effects of anthropogenic land use on the planet are extensive enough to warrant the naming of a new epoch is, in any case, gaining currency everywhere. Since 2008, the Subcommission on Quaternary Stratigraphy's Working Group on the Anthropocene, chaired by stratigrapher Jan Zalasiewicz, has been discussing whether or not to recognize the Anthropocene officially as a new planetary epoch and on what terms. Despite the fact that the Anthropocene has not been officially adopted by the scientific community, there have been explorations of the Anthropocene idea in the mainstream media, and an exhibition entitled "The Anthropocene: The World in Our Hands" at the Deutsches Museum in Munich—currently the largest museum of science and technology in the world—was opened (by Jan Zalasiewicz, among others) in December 2014, indicating the popular imaginative appeal of the term.

The concept of the Anthropocene has already had a marked impact on scholarship in the humanities. Historians and philosophers have endeavoured to grapple with the implications of geological timescales for the human understanding of our own history and future. Dipesh Chakrabarty is one of the most influential thinkers on the implications of the Anthropocene for historiography. His "The Climate of History: Four Theses," published in 2009, continues to provoke discussion amongst humanities scholars (see e.g., Emmett and Lekan 2016). As Chakrabarty lays out at the start of this work, his four theses are based on "the scientific consensus around the proposition that the present crisis of climate change is man-made" (Chakrabarty 2009, 201), and thus are firmly aligned with the central premise of the Anthropocene concept. The "Four Theses" serve in this book as a framework for the exploration of the paradoxes of the Anthropocene.

Chakrabarty aligns the Anthropocene with the Industrial Revolution and the widespread use of gas and oil to drive a new progressive world order. The question of whether the cause of the geological rupture of the Anthropocene should more properly be pinned to the sea change of agriculture, or to the singular force of nuclear fission, goes beyond the scope of this book, but also seems to me to overlook the way that the Anthropocene serves us in the humanities. We cannot know the full extent of what the particular combination of human and more-than-human agencies has unleashed, nor identify a single cause with certainty, but we can think with the concept of the Anthropocene to pick at particular knots in the way that human history, natural processes, and geological forces are entwined. In this book, the knots of the fossil fuel age seem to be particularly worth teasing out, not necessarily because the Anthropocene started here, but because the Industrial Revolution generated a particularly intimate entanglement.

Chakrabarty's first thesis concerns what he calls the "collapse of the age-old distinction between natural history and human history" (2009, 201) and the rise of humanity as a geological, as opposed to a merely biological, agent. "To call ourselves geological agents," Chakrabarty writes, "is to scale up our imagination of the human" (206). This idea shows the Anthropocene as marking a fundamental change in our understanding of what the human category can encompass. The second thesis, "The Idea of the Anthropocene, the New Geological Epoch When Humans Exist as a Geological Force, Severely Qualifies Humanist Histories of Modernity/Globalization" (207), follows on from the first and makes explicit reference to the term Anthropocene. It is also relevant to the literary texts in this study. Here, Chakrabarty explores the conflict between the progressive historical narrative of freedom and enlightenment on the one hand, and on the other the acquisition of destructive geological agency that, as we know only in retrospect, accompanied these developments. Thus, the Anthropocene sets up a counter-narrative to the Enlightenment ideas about nature and progress that stood at the root of many of the transformational water engineering projects on the North Sea coasts during the eighteenth and nineteenth centuries. Freedom for the inhabitants of the lowlands meant freedom from the fear of unpredictable and catastrophic floods, as well as freedom from the poverty and hard labour that dominated life, "the spectre of famine and starvation [that] hung in the air over the heads of our ancestors like the ague in the swamp itself" (Rotherham 2013, 13) in the fens and marshes. A critique of Enlightenment ideals is not the aim; Chakrabarty writes:

> The Anthropocene, one might say, has been an unintended consequence of human choices. But it is also clear that for humans any thought of the way out of our current predicament cannot but refer to the idea of deploying reason in global, collective life.... Logically, then, in the era of the Anthropocene, we need the Enlightenment (that is, reason) even more than in the past.
>
> (210)

The third thesis draws on Chakrabarty's awareness of postcolonial history to lay out another conflict: one between individuals in this new categorization of the human that, collectively, is both a geological force and a globalized, complex, unequal humanity. It raises the notion of *scale* as a principal problem of the Anthropocene; understanding the entangled causes and effects of the new epoch requires a kind of multi-scalar thinking that allows the human individual, human societies, and the human species as an entity to co-exist and to be brought into conversation with one another. In the fourth thesis, Chakrabarty briefly explores the dilemma of thinking in different scales that, for historians, is "probing the limits of historical understanding" (220).

Chakrabarty's careful problematization of the entanglement between humans and nature does not preclude criticism of the term "Anthropocene" by others: there are pitfalls in the concept that need to be acknowledged here. Many have criticized the term for its focus on the dominant human-centred narrative of planetary change. In her 2013 article in *Environmental Humanities*, Eileen Crist argues that the term promotes the idea of a humanity that is unassailably centre-stage, thus lending legitimacy to the anthropogenic activity it seeks to critique:

> Of equal if not greater significance is what this discourse excludes from our range of vision: the possibility of challenging human rule. History's course has carved an ever-widening swath of domination over nature, with both purposeful and inadvertent effects on the biosphere. For the Anthropocene discourse our purposeful effects must be rationalized and sustainably managed, our inadvertent, negative effects need to be technically mitigated—but the historical legacy of human dominion is not up for scrutiny, let alone abolition.
>
> (Crist 2013, 131)

Andreas Malm and Alf Hornborg develop a similar argument in their 2014 article in the *Anthropocene Review* (in the first issue of the newly founded journal), in which they challenge the idea of the Anthropocene as an imposition of natural sciences-led species thinking on the social (and deeply unequal) processes that have caused climate change. In their view, "species-thinking on climate change is conducive to mystification and political paralysis" (Malm and Hornborg 2014, 67). Others see outright danger in a mainstream adoption of the Anthropocene: neo-materialists such as Tim LeCain assert that the concept of the Anthropocene upholds a misleading distinction between human culture and the natural environment:

> As soon as we begin talking about "man as a geological agent" who is taking us into a new "age of humans," we begin to overestimate human power and agency, tending towards a celebratory stance even when the intent is to be critical. Superman, one might observe, can use his super powers for good or evil, but he is still super either way.
>
> (LeCain 2015, 4)

LeCain argues that "humans and their cultures are best understood as the products of their material environment, not its masters" (2) and that the power ascribed to humanity by the term Anthropocene adds unwarranted credence to the advocates of the so-called "Good Anthropocene," in which humans might outwit environmental change by means of massive and vastly ambitious climate-engineering projects. In his recent book *The Matter of History*, LeCain goes on to develop this argument to advocate eloquently for the term "Carbocene" as an alternative term that recognizes the (enduring) dominance of non-human agencies over human narratives of control (LeCain 2018, 311–28).

Donna Haraway has presented the case for multiple names to reflect the different emphases of the new geological epoch, and the scales on which they operate. "I think a big new name, actually more than one name, is warranted," she writes. "Thus, Anthropocene, Plantationocene, and Capitalocene … the Chthulucene.… All the thousand names are too big and too small" (2015, 160). For Haraway, the Anthropocene necessitates plural thinking. Haraway's concerns echo Chakrabarty's problematization of scale; the latter's postcolonialist unease about the collapse of the distinction between oppressed and oppressor—between geological agent and climate-change victim—in the exploitation of the global-ecological category of the human is another critical attitude to the idea of the Anthropocene that is shared by many. Any attempt to gather up all of humanity under a single term can only blur the inequalities and uneven sharing of the powers and benefits of fossil fuel use.

Sharing Haraway's sense that none of the lexical items we have at our disposal is up to the titanic semantic challenge of a planetary epoch, I have nonetheless elected in this book to stick with the Anthropocene as a term that, in its contention, bundles many discourses and ideas into a taut knot. That the terminology remains contentious is not surprising for a concept of such global and *longue-durée* application: geological epochs have hitherto been defined firmly in retrospect, and the very short *durée* of human language with respect to the planet is a markedly limited tool for the task of naming a process that extends into a completely unknowable future. Donna Haraway writes that she sees the Anthropocene as "more a boundary event than an epoch" and adds that:

> The Anthropocene marks severe discontinuities; what comes after will not be like what came before. I think our job is to make the Anthropocene as short/thin as possible, and to cultivate with each other in every way imaginable epochs to come that can replenish refuge.
>
> (2015, 160)

Paraphrasing Haraway, the Anthropocene is a boundary term: it opens the possibility of bigger and multiple words. Its particular anthropocentrism is less troubling in this book, as I am just as concerned with the human imagination of the physical world as I am with the physical world itself. The Anthropocene as a human concept, a new marker in the story of human reflexivity, is a necessary part of a humanities concerned "with the study of intellectual creation and the critique of dominant narratives, myths, and ideologies, and the critical engagement with fundamental questions of meaning, value, responsibility, and purpose in a period of escalating crisis" in Sabine Wilke's words (Wilke 2013, 67). The collision of different time scales—the *longue-durée* of planetary geology, the long human history of weathering storms, and the short human experience of breached dykes—convene in the human understanding of the lowlands; as do human attempts to master the natural world in a setting where natural disasters have been frequent and catastrophic. The Anthropocene concept is not just fruitful as a way of re-reading the North Sea lowlands in literature; the study of literary texts connected to a landscape contributes, in its turn, to a warm-blooded understanding of what the Anthropocene is and can become. Literary texts can encode an awareness of environmental and social inequalities, complex narratives of decline and progress, multiple scales, shifting perceptions of place, and the imagination of the future—in short, they can explore all the paradoxes and contradictions inherent in the idea of the Anthropocene.

In the foreword to their book about climate change and the North Sea coastal cultures, Karsten Reise and Ludwig Fischer write that it is the knowledge of the coming catastrophes that separates this new epoch from the ones that came before; whereas previous generations were used to regular flooding and storm surges, "today we are on the cusp of a rupture" ("stehen wir heute vor einer historischen Zäsur"; Reise and Fischer 2011, 9). We are on the edge, on a metaphorical littoral, of a whole new era. The literary texts analyzed in this book can utilize the liminal moment of the Anthropocene, this moment of discontinuity and reflection, to re-examine the history, present, and future of life in a particular landscape. The concept of the Anthropocene, and its criticism, provides a way of thinking about the impact of the human species on the planet. The North Sea lowlands, with their shared history of human engineering and ecological vulnerability, provide a theatre in which the ideas and implications of a transition into the Anthropocene can be played out.

About this book

This book looks comparatively at literary texts set in a geographical area spanning four countries. In Europe, environmental historians have led the way in studies of knowledge transfer that are transnational, breaking out of the determinedly nationalist histories that grew up through the eighteenth and nineteenth centuries, and looking both at the *longue-durée* and across national borders to better understand the interactions between humans and their landscapes. Studies such as Fernand Braudel's *La Méditerranée et le monde méditerranéen à l'époque de Philippe II* (1949) and Alfred Crosby's *The Columbian Exchange: Biological and Cultural Consequences of 1492* (1972), for example, were influential in their analysis of transnational processes and their rejection of deterministic nationalist narratives. The sub-field of water history has championed a drainage-basin approach to studying landscapes, using the underlying cyclical systems of groundwater and river flow that provide ecological boundaries between areas (which sometimes run counter to contingent political borders), to decentre the national and cultural histories and reveal human entanglements with the more-than-human environment (see for example Marc Cioc's 2002 book *The Rhine: An Eco-biography*, which follows the Rhine across several countries; and Bo Fritzbøger's *Vandets Veje* from 2009 for its history of a river basin over 350 years).

The approach taken by this book, in setting the area for study as the drainage basin of the lower North Sea, renders *Shifting Sands* a necessarily transnational affair; the various European literatures studied here have, for the most part, been studied discretely, in the celebratory literary traditions that grew up as part of the nation-building enterprise in the nineteenth century. Area studies, such as Scandinavian Studies and North American Studies, have developed limited transnationality in their approach to literatures (as well as being comparative in the disciplines they draw on, which include cultural studies, literature, history, and linguistics), but have not sought to do away with geopolitical borders in doing so; the comparative literature that grew up in Europe as a political response to nationalism pre-1945 is built on the idea of comparing literature between literary traditions, rather than within them, but its strength has been its focus on intellectual and philosophical transfers, rather than thematic ones (Hutchinson 2018, esp. 23–5; Spivak 2003, 1–22; see also Lubrich 2006, 47–67 on Germany, and Lernout 2006, 37–46 on the Low Countries).

As environmental historians are wont to reiterate, the environment doesn't stop at national borders, and thus comparative scholars, with their multilingual and multicultural expertise, should be at the forefront of ecocritism. Yet the discipline of comparative literature has nevertheless been surprisingly slow to address pressing environmental issues. Likewise, postcolonial literary studies remain tied to the legacies of empire, expansion, and power that grew out of configurations of nationalism, even as transnational ecocriticism is working to counter this with a truly transnational approach to global problems.[2]

The primary texts studied in this book are linked by their use of the landscapes of the North Sea lowlands to express and explore particular ideas about humans and nature in the Anthropocene. In each of them, the landscape goes beyond setting, and is an active driver of the plot. The use of landscape in the selected works falls within Lawrence Buell's conception of environmental text, outlined in his 1995 book on Thoreau and American nature writing: Buell gives four criteria (which he describes as a non-exhaustive "rough checklist of some of the ingredients that might be said to comprise an environmentally oriented work") for identifying an environmental text:

> 1 The nonhuman environment is present not merely as a framing device but as a presence that begins to suggest that human history is implicated in natural history. [...]
> 2 The human interest is not understood to be the only legitimate interest. [...]
> 3 Human accountability to the environment is part of the text's ethical orientation. [...]
> 4 Some sense of the environment as a process rather than a constant or a given is at least implicit in the text. [...]
>
> (Buell 1995, 7)

These criteria, to a greater or lesser extent, describe the representation of the landscape of the North Sea coasts in the literary texts looked at in this book.

Buell stipulates that in his environmental texts, the landscape is intrinsic rather than peripheral to the text; it has its own history and future, independent of the history and future of the human characters; the text examines questions about the human habitation of the space; and the natural environment is shown to be in flux, rather than completely static. Employing these criteria enables me to draw together texts that range widely across genre and language, making a strong case for a broad definition of what an environmental text can be. Literary texts that seem to focus on a human story can, on close reading, be shown to have a strong environmental narrative as well; in some cases, the marketing of books plays down the natural setting in favour of human interest and, in many cases, a critical tradition that has focused on the inner life of humans has overlooked the environmental concerns that underwrite the human characters. This broad definition of an environmental text is necessary in order to discern trends and motifs common to the environmental imagination of particular places.

Beyond these key criteria and the focus on the specific landscape of the North Sea lowlands, my choice of texts in this book was not constrained by genre, language, or time period (beyond the interest in modern literature): the aim has been to reframe the literary texts within a topographical and transnational cultural context. The act of examining a collection of texts in their relation to each other purely via their engagement with a particular geographical environment destabilizes the linear, historical, and often also canonical context that makes up much of the work of literary scholarship, allowing both established and newer or marginal works to benefit from a fresh perspective and environmental focus. As a whole, the selection of literary works presents a diffused and variegated picture of the literary Anthropocene, as it is played out in a particular landscape; across the text corpus, there are both discernible shifts and surprising commonalities in the way that humans perceive the natural environment around the fragile coasts of the North Sea.

Thus, we have novels, prose, short stories, and poetry in different languages, creating a textual and linguistic mosaic that shows convincingly how particular imaginations of the North Sea coast are held in common. The texts represent quite different historical and cultural contexts; they are written in five languages—Danish, Dutch, English, German, and Norwegian—across quite a long time period (but one fairly congruent with Stoermer and Crutzens' own definition of the start of the Anthropocene, as discussed above); the earliest text, Hans Christian Andersen's *De to Baronesser* was published in 1848; Jim Shepard's story "The Netherlands Lives with Water" first appeared in 2010. The emphasis falls more heavily on recent works—the burgeoning interest in the current crisis of climate change seems to have led to a great engagement with the environment in literary form as defined in Buell's criteria in recent decades. The motifs and themes that I have identified in this book—the dynamism of natural systems, for example, the instrumentalization of North Sea risk culture, and the imagination of floods past and present—point to an interplay between the understanding of physical place and its imagination in literary form.

Given the pressing need to address the critical environmental issues across a broad front, there is a need in humanities scholarship for this kind of comparative study. It identifies texts that are environmental in a wider sense, and across a wide area; Adam Trexler writes in his recent study of English-language novels in *Anthropocene Fictions* that "it remains difficult to locate such novels [climate change novels in languages other than English], although there is a vital need for the cross-cultural insight they could provide and a commensurate need for scholarship by critics specializing in other languages" (Trexler 2015, 10). One or two of the texts in this book (*Der Schimmelreiter* by Theodor Storm, and W.G. Sebald's *Die Ringe des Saturn: Eine englische Wallfahrt*, for example), are canonical, or at least critically acclaimed, within their respective national literary traditions (and, in the case of Sebald, within a neighbouring literary tradition almost more than his own); others are far less well-known, or are simply too recently published to have established themselves firmly within scholarly discourse. The fact that English-language texts dominate overall in this selection, besides reflecting the difference in population size, perhaps also signals a greater engagement with the fast-eroding East Anglian littoral on the part of British writers; the fact that Dutch texts are fewer may likewise in part be a figuration of a general sense that "real" nature is absent from the quintessentially Dutch lowland landscape, as the academic and writer Isabel Hoving has indicated (Hoving 2010).

The fact that a range of literary forms are included means that the comparisons necessarily concentrate primarily on thematic elements, although the formal elements of structure and style remain important for each close reading. Thematic overlaps are found across such a range of texts—from children's novels to non-fiction to avant-garde poetry—which speaks for the coherence of the imaginary landscape as a site where particular questions about humans in nature can be broached. Of course, the breadth of texts addressed in the book is buttressed by an even greater corpus of primary texts that were not ultimately included in the close analysis in each chapter. This book unfolds in six chapters that function as case studies: each chapter has one or two works at its heart chosen to represent particular themes or modes of reflection on the landscape. The texts chosen were the ones that best met Buell's criteria but the themes were tracked across a large corpus of other texts and thus each chapter can be seen as examining a particular *kind* of landscape perception.

For example, many texts besides Storm's *Der Schimmelreiter* (*The Rider on the White Horse*) explored the lowlands as the site of uncanny or mysterious events, or as spaces of lawlessness; as individual texts they conformed only to a limited extent to Buell's criteria for environmental texts, but their reading certainly informed a sense of the history of this landscape as an imaginary space with fictional uses. Wilkie Collins set *The Moonstone* (1868, generally seen as England's first detective novel) on the East Anglian coast, and Erskine Childers, whose 1903 work *The Riddle of the Sands* is widely regarded as the first ever spy novel, used the transnational potential of the North Sea basin and the difficulty of navigating the shallow sands around the coasts by boat as a plot device. This text found echoes in Arthur Ransome's sailing adventure stories for young readers, such as *The Coot Club*, *We Didn't Mean to Go to Sea*, *Secret Water*, and *The Big Six*, which were first published between 1934 and 1940 and are all set in the treacherous coastal waters around the North Sea basin. Dorothy Sayers also exploited the eerie fenland landscape and the dramatic potential of flooding and water management for her 1934 murder mystery *The Nine Tailors*. These texts have contemporary descendants in suspense texts set on the North Sea coasts, such as those by Ann Cleeve, Håkan Nesser, Elke Bergsma, and Klaus-Peter Wolf.

Likewise, the migrating birds that are important figures in Kjersti Vik's novel *Mandø*, and which have inspired many of the nature writers discussed in Chapter 4, are also found in other texts set on the North Sea. The ecological subtext of contemporary works such as Christiane Neudecker's coming-of-age *Sommernovelle* (2015), in which two young activists are divested of their environmental innocence on the island of Sylt, and Uwe Timm's *Vogelweide* (2013), in which the narrator recounts his intertextual story of adultery and loss from a German "Hallig" where he lives alone, employed to count the birds that nest there, is hard to miss. There are historical progenitors here too. Two earlier novels that remain enduringly popular, Paul Gallico's *The Snow Goose* (1941) and Meindert DeJong's *The Wheel on the School* (first published in 1954) testify to the appreciation of the importance of birds to communities in the North Sea lowlands; one of Arthur Ransome's adventure stories set in East Anglia (*The Coot Club*, 1934) is also concerned with the preservation and appreciation of birdlife in Norfolk. An understanding of the way that birds have been part of the perception of this landscape in other literary texts informs my reading of *Mandø*.

The material of the land itself is a thread running through literary texts besides those in this book. The mud and sodden land of Arthur van Schendel's *De Waterman* and the migrating sands of Hans Christian Andersen's Oland in *De to Baronesser* (both explored in Chapter 1) are critical features of the landscape; in Hans Christian Andersen's short story "En historie fra Klitterne" (1859), the agency of sand is also crucial to the plot. Robert Macfarlane's prose text "Silt" captures the sensations of walking out on the eponymous ground of the North Sea shallows at low tide and reflects on the mutability of the landscape. A longer text concerned likewise with silt is Graham Swift's 1983 novel *Waterland*. This work is roughly contemporary with W.G. Sebald's *Die Ringe des Saturn* and, like the latter, deals with the shifting grounds of rivers and watercourses and the shifting and unstable narratives of history. Swift sets his novel in the East Anglian Fens, and writes of the "slow, sly, insinuating agency" of silt (Swift 1983, 8). Like Sebald, Swift uses the landscape as a springboard for the examination of human and natural histories; Sebald's interest in silkworms and herring is mirrored in Graham Swift's exploration of the life cycle of the eel, as pointed out by scholar Bernhard Malkmus (Malkmus 2017).

Literary explorations of environmental apocalypse have frequently invoked flooding, and the trope remains a powerful figure in climate change fiction (Rigby 2015, 84–111; Trexler 2015, 75–118). The North Sea basin, site of historic floods, provides the location for some dystopian perspectives on the coastal areas. Jon McGregor's short story collection *This Isn't the Sort of Thing That Happens to Someone Like You* (2012) contains intimations of a coming catastrophe, whether by flood or by human malevolence, in the Fens of East Anglia. In 2015, the Australian novelist James Bradley published *Clade*, a futuristic account of climate change and disaster, in which the globetrotting characters move across continents, but the storm surge and resulting large-scale migration of people start on the eastern coast of England. Isabel Hoving's 2015 novel *De een na laatste dood van het meisje Capone* also takes up in a drowned Holland and Germany in the year 2099, and could be examined alongside Øyvind Rimbereid's future scenario in "Solaris korrigert," which is the subject of Chapter 6. Eva Moraal's novel *Overstroomd*, originally published in 2012, is likewise set in the Netherlands following a catastrophic flood that has divided the nation into wet and dry sections inhabited by different classes of society.

The subject of human hostilities and warfare in the twentieth century is also found in specific constellations around the North Sea. Notable amongst German-language texts is Siegfried Lenz's contribution to the German literature of "Vergangenheitsbewältigung" (attempts to work through the trauma of the National Socialist past), in particular his novel *Deutschstunde*, based on the fate of the celebrated expressionist painter Emil Nolde under the Nazis. Central to Lenz's work is the Frisian coast as a space on the margins of the German nation. Another German author to explore the aftermath of war and fascism on both the North Sea and Baltic Sea coasts of Germany is Jochen Missfeldt, for example in *Solsbüll* (1989). Across the sea in England, W.G. Sebald's *Die Ringe des Saturn* touches on the bombing missions that started out from the Fens and swept across the North Sea to Germany, as does Graham Swift in *Waterland* and many of the English nature writers explored in Chapter 4, in particular Helen MacDonald in *H is for Hawk*. Several poems in Blake Morrison's 2015 collection *Shingle Street* allude to the legend of a failed German invasion on the East Anglian coast and to the evacuation of coastal inhabitants in the 1940s.

Finally, there are texts strongly associated with the North Sea coasts that did not fit easily into any of the broader themes explored here. Karen Blixen's "The Deluge at Norderney" (published in English in 1934 and in Danish as *Syndfloden over Norderney* in 1935 under the pseudonym Isak Dinesen) is one such text; the setting of the Frisian island in the early years of North Sea tourism and the sudden appearance of a dramatic storm surge fulfil the criteria Lawrence Buell sets out for an environmental text, but the challenging novella bears little comparison to other works set in this landscape. The Nordeney invoked by Blixen is neither windswept sublime nor muddy reality, but something in between: "Nordeney is not an obvious place for a seaside resort: it is, in fact, only made one by the romantic imagination," Rachel Trousdale notes (Trousdale 2002,

207). Nordeney is a real place, but is also constructed by the strange, almost grotesque characters in this story.

The chapters of this book look both backwards and forwards from a crucial "knot" in the coming into being of the Anthropocene, that of the idea of human progress and freedom from environmental limits that was brought about primarily by new technological developments. In the opening chapter, "Against the tide: living with the North Sea," I look at literary portrayals of the risk culture of the North Sea lowlands. Novels by the Danish author Hans Christian Andersen (*De to Baronesser*, 1848) and the Dutch writer Arthur van Schendel (*De Waterman*, 1933), written 100 years apart, address the way that our understanding of storm floods as regular occurrences in this landscape is mediated by cultural or religious worldviews, particularly with regard to the rise of the nation-state and narratives of technical progress. The final part of this chapter looks briefly at a range of literary responses to the 1953 storm surge, detecting a postmodern cynicism towards any common flood narrative, whether spiritual or scientific, in the literature of the later twentieth and early twenty-first centuries.

In the second chapter, "Conquest and control: engineering the Anthropocene on the North Sea," the role of dyke and drainage technology and of the limits of human power against the North Sea tides is explored through a close reading of two texts: Theodor Storm's *Der Schimmelreiter* (*The Rider on the White Horse*, 1888) and a 2010 short story by the American writer Jim Shepard entitled "The Netherlands Lives With Water." Storm's famous historical novella shows the process of land reclamation in the early eighteenth century as fraught with the tension between Enlightenment and progress on the one hand and superstition and fear on the other; while many have read the novel as the tragic tale of a romantic prophet-engineer, recent ecocritical scholarship has noted a narrative of unintended consequences that highlights the Enlightenment's colonialist approach to nature. The second text, Shepard's "The Netherlands Lives with Water," is set in a future Rotterdam in which the water flows underground, rendered invisible by technological advances; the unintended consequences of anthropogenic climate change are rising sea levels and increased rainfall that spell disaster for the low-lying city and its inhabitants. The two texts are linked not just thematically by their exploration of human attempts to control the hazardous environment of the lowland coasts, but also by their use of narrative strategies that question the idea of the worldview of "progress" behind land reclamation and the protection of urban settlements.

The third chapter is "A sense of place in the Anthropocene" and, alone in this book, it focuses on a single work, W.G. Sebald's *Die Ringe des Saturn: Eine englische Wallfahrt* (translated and published in English as *The Rings of Saturn* in 1997). Sebald's book situates the English lowlands at the heart of a complex of global interactions and timescales. As Adam Trexler points out, "American and British environmentalists have often argued that a sense of place is central to the project of conservation" (Trexler 2015, 75) and this work combines an exploration of the interconnectedness of a specific coastal landscape—a backwater—with histories from all corners of the globe. Emphasizing this resolution of the perceived divide between place and space and his telescoping of chronologies, the chapter concludes with an appraisal of this book as an account of the transition to the Anthropocene, drawing on Dipesh Chakrabarty's exposition of the paradoxes of perspective and scale. In Sebald's text, the project of human Enlightenment is hopelessly entangled with the darkness and savagery of colonialism and ecological exploitation, and the eroding coasts of the North Sea, slowly migrating westwards, are both a site and a symbol of human progress and decline.

The popular vein of contemporary writing labelled as "new nature writing" has been inspired in no small part by Sebald's *Rings of Saturn*, and the fourth chapter, "Landscape as palimpsest: East Anglia in British 'new nature writing,'" examines contemporary examples of this subgenre concerned, like Sebald, with the coast and fens of East Anglia. These hybrid texts, which mix aesthetic and affective strategies with descriptions of botany, geology, and history, present an interplay of perspectives from the subjective solitary human to the geological and the planetary, thus continuing Sebald's project of bringing different perspectives and timescales into conversation. In this chapter, I explore the genre using examples of texts written about the East Anglian lowlands by Robert Macfarlane, Richard Mabey, Helen Macdonald, and Mark Cocker. A close reading of two short prose texts by the unofficial British figurehead of new nature writing, Robert Macfarlane, show how this kind of text is responding to a threatened landscape by attempting to give authentic expression to the human relationship with the natural world. The fragility of the East Anglian landscapes renders these a theatre for the complex ecologies of the Anthropocene.

The penultimate, fifth, chapter is concerned with contemporary fiction set on the margins of the North Sea. In "Causeways to the past: Anthropocene and memory in contemporary novels" the two primary texts are novels set on the very edge of the water—the Danish island of Mandø in Kjersti Vik's novel of the same name and an almost abandoned stretch of English coast in Meg Rosoff's *What I Was*. The chapter takes as its premise the crossing and attempted crossings of half-submerged causeways as a figuration of the fragile bridge between almost extinct coastal natures and the modern and largely urban culture that has become dominant during what we think of as the Anthropocene; and also between the natural world of the past and the possibilities for human environments to come. Both novels invoke the history of the landscape as a way of understanding the future.

The sixth and final chapter, "Under the North Sea: 'petrospectral' futures," takes as its starting point a short story by the British New Weird author China Miéville, and moves from there to a close reading of the complex long poem "Solaris korrigert" by the Norwegian writer Øyvind Rimbereid. This chapter shows the new visibility of oil in the understanding of the imaginary space of the North Sea. In these two unsettling works, humans are simultaneously technological masters and a biological species in retreat: following on from some of the themes introduced in Chapter 5, in which Darwinian ideas about the adaptation of organisms to specific ecosystems are discussed, both works suggest an entanglement of human and material worlds that requires new levels of adaptation. Meanwhile, in Rimbereid's imagination of the future, the region has coalesced around oil to produce a hybrid language drawn from all sides of the North Sea basin.

My approach to this book is defined by its breadth in respect of the diversity of primary sources. As a work aiming to contribute to the Environmental Humanities, it exploits the latter's openness towards different disciplines. Environmental history, philosophy, ecology, geology, geography, and natural history are all woven into the theoretical framework for the literary texts examined here, and with each of these disciplines I have had recourse to a large and growing body of secondary literature. This kind of approach draws its strengths from its breadth; I have focused on the explicit and implicit *connections* between quite disparate texts, between different historical moments and genres. I chose to structure this project thematically to make the most of these broad connections. The intersections between the texts are manifold; organizing the texts by genre would have meant missing national-cultural and chronological overlaps; organizing by chronology would have glossed over some of the thematic similarities between works whose dates of composition were far apart; organizing along national or linguistic lines would have obscured many of the most striking common features shared between works written in different languages. In short, a more conventional approach would have been unable to do justice to the richness of our shared environmental imaginary. Literary texts and literary analysis are important if we are to understand the breadth and depth of human cultural complicity in, and responses to, the Anthropocene.

At a time in which our understanding of our place in nature is frequently called into question, and in which engagement in local and global ecologies has taken on a pitch of urgency, understanding how sense of place is constructed and maintained within the environmental imagination, how human societies deal with the risks of coastal engineering, with "commons" such as a marine ecosystem like the North Sea, and what the future holds in the Age of Humans, cannot be more pressing. The literary imagination of landscape is a site for critical and emotional responses to its history, present, and future. This liminal and fluctuating environment is vulnerable to anthropogenic climate change now, but it has always been exposed to instability and transformation. As a prism through which we can think through these past and future vulnerabilities, the North Sea lowlands can serve as a bellwether for the whole of the planet and the whole of its humanity.

Notes

1 According to the Ramsar website (www.ramsar.org/country-profiles) (last accessed 2 July 2018), the UK has 174 such protected sites, the Netherlands 54, Denmark 43, and Germany 34.

2 The DFG Research Network at the University of Augsburg entitled "Environmental Crisis and the Transnational Imagination" and headed by Timo Müller identifies the inadequacy of national solutions to a global problem and aims "to ask in how far the environmental crisis has put the cultural concept of nation into question already, and to identify alternative approaches," noting the importance of cultural narratives in instituting real change. www.philhist.uni-augsburg.de/de/lehrstuehle/anglistik/network/ (last accessed 3 July 2018).

References

Agnew, John A. 2011. "Space and Place." In *The Sage Handbook of Geographical Knowledge*, edited by John A. Agnew and David N. Livingstone, 316–30. London: Sage.

Andersen, Hans Christian. 1860. "En Historie fra Klitterne." In *Nye Eventyr og Historier af H. C. Andersen. Fjerde Samling*, 35–57. Copenhagen: Reitzels Forlag.

Andersen, Hans Christian. 1866. *De to Baronesser*: Original Roman i tre Dele. Copenhagen: Reitzels Forlag.

Appadurai, Arjun. 1990. "Disjuncture and Difference in the Global Cultural Economy." *Theory Culture Society* 7 (2–3): 295–310.

Bankoff, Greg. 2013. "The 'English Lowlands' and the North Sea Basin System: A History of Shared Risk." *Environment and History* 1: 3–37.

Barkham, Patrick. 2014. "Should Coastal Britain Surrender to the Tides?" *Guardian*, 7 February 2014. Accessed 24 March 2016. www.theguardian.com/environment/2014/feb/07/should-coastal-britain-surrender-to-tide.

Barkham, Patrick. 2015. "This Sinking Isle: The Homeowners Battling Coastal Erosion" *Guardian*, 2 April 2015. Accessed 2 July 2018. www.theguardian.com/news/2015/apr/02/sinking-isle-coastal-erosion-east-anglia-environment.

Beck, Ulrich, Wolfgang Bonss, and Christoph Lau. 2003. "The Theory of Reflexive Modernization: Problematic, Hypotheses and Research Programme." *Theory Culture Society* 20 (1): 1–33.

Blackbourn, David. 2007. *The Conquest of Nature: Water, Landscape and the Making of Modern Germany*. London: Pimlico.

Blixen, Karen. 1961. "Syndfloden over Nordeney." In *Syv fantastiske Fortællinger*, 165–262. Copenhagen: Gyldendal.

Blixen, Karen (Isak Dinesen). 1979. "The Deluge at Nordeney." In *Seven Gothic Tales*, 117–82. St Albans: Triad.

Bradley, James. 2017. *Clade*. London: Titan Books.

Braudel, Fernand. 1972. *The Mediterranean and the Mediterranean World in the Age of Philipp II*. Translated by Sian Reynolds. London: Collins.

Buell, Lawrence. 1995. *The Environmental Imagination: Thoreau, Nature Writing, and the Formation of American Culture*. Cambridge, MA: Harvard University Press.

Carolyn Merchant. 1980. *The Death of Nature: Women, Ecology, and the Scientific Revolution*. San Francisco, CA: Harper & Rowe.

Chakrabarty, Dipesh. 2009. "The Climate of History: Four Theses." *Critical Enquiry* 35 (2): 197–222.

Childers, Erskine. 2007. *The Riddle of the Sands*. London: Penguin.

Cioc, Marc. 2002. *The Rhine: An Eco-biography, 1815–2000*. Seattle: University of Washington Press.

Collins, Wilkie. 1868. *The Moonstone*. London: The Tinsley Brothers.

Crist, Eileen. 2013. "On the Poverty of Our Nomenclature." *Environmental Humanities* 3: 129–47.

Crosby, Alfred. 1972. *The Columbian Exchange: Biological and Cultural Consequences of 1492*. Westport, CT: Greenwood.

Crutzen, Paul J. and Eugene F. Stoermer. 2000. "The Anthropocene." *IGBP Newsletter* 41: 17–18.

DeJong, Meindert. 1988. *The Wheel on the School*. London: Puffin.

Emmett, Robert and Thomas Lekan. 2016. "Whose Anthropocene? Revisiting Dipesh Chakrabarty's 'Four Theses.'" *RCC Perspectives* 2.

Fernandez-Bilbao, Amalia. 2012. "Adaptation to Climate Change by Coastal Communities." In *Vulnerability, Risks, and Complexity: Impacts of Global Change on Human Habitats. Advances in People-Environment Studies* 3, edited by Sigrun Kabisch, Anna Kunath, Petra Schweizer-Ries, and Annett Steinführer, 53–65. Cambridge, MA: Hogrefe.

Fischer, Ludwig and Karsten Reise. 2011. "Vorwort zum Buch." *In Küstenmentalität und Klimawandel: Küstenwandel als kulturelle und soziale Herausforderung*, edited by Ludwig Fischer and Karsten Reise. Munich: Oekom, 7-8.

Fritzbøger, Bo. 2009. *Vandets Veje: Skjern Ås miljøhistorie gennem 350 år*. Copenhagen: Gyldendal.

Gaffney, Vincent, Simon Fitch, and David Smith. 2009. *Europe's Lost World: The Rediscovery of Doggerland*. York: Council for British Archaeology.

Gallico, Paul. 1941. *The Snow Goose*. London: M. Joseph.

Gerdes, Rüdiger. 2011. "Meeresspiegelanstieg—Prognosen und Szenarien." In *Küstenmentalität und Klimawandel: Küstenwandel als kulturelle und soziale Herausforderung*, edited by Ludwig Fischer and Karsten Reise, 117–30. Munich: Oekom.

Gerritsen, Herman. 2005. "What Happened in 1953? The Big Flood in the Netherlands in Retrospect." *Philosophical Transactions of the Royal Society* A 363: 1271–91.

Giblett, Rodney James. 1996. *Postmodern Wetlands: Culture, History, Ecology*. Edinburgh: Edinburgh University Press.

Hall, Alexander. 2013. "The North Sea Flood of 1953." Environment & Society Portal, *Arcadia* 5. Rachel Carson Center for Environment and Society. Accessed 2 July 2018. www.environmentandsociety.org/node/5181.

Haraway, Donna. 2015. "Anthropocene, Capitalocene, Plantationocene, Chthulucene: Making Kin." *Environmental Humanities* 6: 159–65.

Haraway, Donna. 2016. *Staying with the Trouble: Making Kin in the Chthulucene. Experimental Futures*. Durham, NC: Duke University Press.

Hasse, Jürgen. 2011. "Zur mythischen Funktion deklarierter Natur-Landschaften: Das Beispiel des 'Weltnaturerbes' Wattenmeer." In *Küstenmentalität und Klimawandel: Küstenwandel als kulturelle und soziale Herausforderung*, edited by Ludwig Fischer and Karsten Reise, 97–113. Munich: Oekom.

Hillmer, Angelika. 2013. "Niederländer geben Flüssen mehr Raum." *Die Welt*, 19 July 2013. Accessed 2 July 2018. www.welt.de/wissenschaft/umwelt/article118194218/Niederlaender-geben-Fluessen-mehr-Raum.html.

Hoving, Isabel. 2010. "On the Absence of Nature: Writing on Nature and Ecocriticism in the Netherlands." *Ecozon@* 1: 167-73.

Hoving, Isabel. 2015. *De een na laatste dood van het meisje Capone*. Amsterdam: Querido.

Hutchinson, Ben. 2018. *Comparative Literature: A Very Short Introduction*. Oxford: Oxford University Press.

Kelly, Matthew. 2018. "The Thames Barrier: Climate Change, Shipping, and the Transition to a New Envirotechnical Regime." In *Histories of Technology, the Environment, and Modern Britain*, edited by Jon Agar and Jacob Ward, 206–29. London: UCL Press.
LeCain, Timothy J. 2015. "Against the Anthropocene: A Neo-Materialist Perspective." *International Journal for History, Culture and Modernity* 3 (1): 1–28.
LeCain, Timothy J. 2018. *The Matter of History: How Things Create the Past*. Cambridge: Cambridge University Press.
Lenz Siegfried. 1968. *Deutschstunde*. Hamburg, Hoffmann and Campe.
Lernout, Geert. 2006. "Comparative Literature in the Low Countries." *Comparative Critical Studies* 3 (1–2): 37–46.
Lubrich, Oliver. 2006. "Comparative Literature—In, From and Beyond Germany." *Comparative Critical Studies* 3 (1–2): 47–67.
Maas, Barbara. 2013. "Hier entsteht Schleswig-Holsteins höchster Deich." *Flensburger Tageblatt*, 5 October 2013. Accessed 2 July 2018. www.shz.de/schleswig-holstein/panorama/hier-entsteht-schleswig-holsteins-hoechster-deich-id3782576.
Mabey, Richard. 2005. *Nature Cure*. London: Chatto & Windus.
Macdonald, Helen. 2014. *H is for Hawk*. London: Jonathan Cape.
Macfarlane, Robert. 2012. "Silt." *Granta* 119 (May): 41–60.
Malkmus, Bernhard. 2017. "Diffuse Dwelling: Landscape in Graham Swift and W.G. Sebald." In *Readings in the Anthropocene: The Environmental Humanities, German Studies, and Beyond*, edited by Sabine Wilke, 263–95. Rochester, NY: Bloomsbury.
Malm, Andreas and Alf Hornborg. 2014. "The Geology of Mankind? A Critique of the Anthropocene Narrative." *Anthropocene Review* 1 (1): 62–9.
McGregor, Jon. 2012. *This Isn't The Sort of Thing That Happens to Someone Like You*. London, Bloomsbury.
Miéville, China. 2015. "Covehithe." In *Three Moments of an Explosion: Stories*, 337–51. New York: Del Rey.
Missfeldt, Jochen. 1989. *Solsbüll*. Ebenhausen/München: Langewiesche-Brandt.
Moraal, Eva. 2015. *Overstroomd*. Rotterdam: Lemniscaat.
Morley, George. 1968. *The North Sea*. London: Muller.
Morrison, Blake. 2015. *Shingle Street*. London: Chatto & Windus.
Neudecker, Christiane. 2015. *Sommernovelle*. Luchterhand. Muenchen.
Parker, Rowland. 1978. *Men of Dunwich: The Story of a Vanished Town*. London: Collins.
Pott, Richard. 2003. Die Nordsee: *Eine Natur- und Kulturgeschichte*. Munich: C.H. Beck.
Pye, Michael. 2015. *The Edge of the World: A Cultural History of the North Sea and the Transformation of Europe*. New York: Pegasus.
Ransome, Arthur. 2001. *Secret Water*. London: Red Fox.
Ransome, Arthur. 2001. *The Big Six*. London: Red Fox.
Ransome, Arthur. 2001. *Coot Club*. London: Red Fox.
Ransome, Arthur. 2013. *We Didn't Mean to Go to Sea*. London: Red Fox.
Rieken, Bernd. 2005. *Nordsee als Mordsee: Sturmfluten und ihre Bedeutung für die Mentalitätsgeschichte der Friesen*. Münster: Waxmann.
Rigby, Kate. 2015. *Dancing with Disaster: Environmental Histories, Narratives, and Ethics for Perilous Times*. Charlottesville: University of Virginia Press.
Rimbereid, Øyvind. 2005. "Solaris korrigert." In *Seine topografier/Trådreiser/Solaris korrigert*, 9–45. Trondheim: Gyldendal.
Rosoff, Meg. 2007. *What I Was*. New York: Viking.
Rotherham, Ian. 2013. *The Lost Fens: England's Greatest Ecological Disaster*. Stroud: The History Press.

Sayers, Dorothy. 1977. *The Nine Tailors*. London: New English Library.

Schama, Simon. 1995. *Landscape and Memory*. New York: Knopf.

Sebald, W.G. 1997. *Die Ringe des Saturn: Eine englische Wallfahrt*. Frankfurt am Main: Fischer.

Shepard, Jim. 2011. "The Netherlands Lives with Water." In *You Think That's Bad: Stories*, 38–66. New York: A.A. Knopf.

Spivak, Gayatri. 2003. *Death of a Discipline*. New York: Columbia University Press.

Steffen, Will. 2013. "Commentary: Paul J. Crutzen and Eugene F. Stoermer, 'The Anthropocene.'" In *The Future of Nature*, edited by Libby Robin, Sverker Sörlin, and Paul Warde, 486–90. New Haven, CT: Yale University Press.

Storm, Theodor. 1888. *Der Schimmelreiter*. Berlin: Verlag der Gebrüder Paetel.

Swift, Graham. 1983. *Waterland*. London: Picador.

Timm, Uwe. 2013. *Vogelweide*. Cologne: Kiepenhauer & Witsch.

Trexler, Adam. 2015. *Anthropocene Fictions: The Novel in a Time of Climate Change*. Charlottesville: University of Virginia Press.

Trousdale, Rachel. 2002. "Self-invention in Isak Dinesen's 'The Deluge at Norderney.'" *Scandinavian Studies* 74 (2): 205–22.

van Schendel, Arthur. 1940. *De Waterman*. Rotterdam: Nijgh & Van Ditmar.

Vik, Kjersti. 2009. *Mandø*. Oslo: Aschehoug.

Ward, Ingrid, Piers Larcombe, and Malcolm Lillie. 2006. "The Dating of Doggerland—Post-glacial Geochronology of the Southern North Sea." *Environmental Archaeology: The Journal of Human Palaeoecology* 11 (2): 207–18.

Weston, Daniel. 2011. "The Spatial Supplement: Landscape and Perspective in W.G. Sebald's *The Rings of Saturn*." *Cultural Geographies* 18 (2): 171–86.

Wilke, Sabine. 2013. "Anthropocenic Poetics: Ethics and Aesthetics in a New Geological Age." In *Anthropocene: Envisioning the Future of the Age of Humans*, edited by Helmuth Trischler. *RCC Perspectives* 3, 67–74.

Wöbse, Anna-Katharina. 2013. "Knechtsand—A Site of Memory in Flux." In "Environment and Memory." Special issue, *Global Environment* 2: 160–83.

1 Against the tide

Living with the North Sea

A legendary event recounted by the medieval chronicler Henry of Huntingdon tells us that King Canute (Cnut), ruler of the North Sea Empire in the late tenth and early eleventh centuries, had his advisors carry him on his throne to the seashore so that he could command the tide to turn back. The tide did not turn, but kept on coming: the power of kings was not greater than the power of God or nature; all men, however exalted, were part of a dependent relationship with the natural world around them (Huntingdon 1753). The veracity of this story is less important than the fact it is still often invoked (Westcott 2011): the longevity of this metaphor shows the power of this imagery—humans on the edge of the shore, at the limits of their agency—on our imaginations.

On the North Sea coasts, regional identities have long been built around proximity to the sea and the power of the tides, and the metaphorical richness of communities living "on the edge" between land and sea has been used to convey ideas about national character and construct national identities. Littoral ways of life and adaptive strategies could easily be extended to serve as metaphors for the successes and failures of human culture more generally, and the boundary between land and sea provides a clear spatial representation of the boundary between culture and nature. Literary texts have the ability both to exploit the figurative potential of this littoral world and provide critical reflection on it.

The perception of the North Sea coast as a hazardous landscape is integral to how it is evoked in literary form. Memories of the great storm surges that have affected communities around the coasts at different times are a key part of this sense of a North Sea history: the current shape of the coast is a direct product of many of these extreme events, and the names of the great floods are remembered in communities along the shores. Understanding of these storm surges has shifted over time, from them being interpreted as divine punishment for human wrongdoing, to natural disasters that could be explained in scientific terms, and, latterly, to symptoms of global environmental crisis, although these viewpoints have also existed alongside one another to a certain extent, and still overlap and contradict each other today (see e.g., Leydesdorff 2000 on the religious responses to the 1953 flood in the Netherlands; for contemporary discussions on the causes of climate change, see Hulme 2009). Historians of risk and disaster studies have analysed patterns and strategies in the way that extreme events are remembered, and how memories of such extreme events aid the management and mitigation of future floods (Bankoff 2013, 8; Mauelshagen 2007; Kempe 2007). The recurrence of these large-scale flooding events is an important part of North Sea cultures: "Repetition ... becomes a key concept in historical research on disasters. It is the link between past experiences and future societies. Prevention and management, because they require prediction, rely on the future of the past" as Mauelshagen (2009, 45) puts it. The act of remembering or memorializing flood disasters, or indeed forgetting or suppressing memories, is an engagement with future successes or failures, and thus takes on a particular role in political and cultural discourse.

Literary explorations of the North Sea floods are an important part of this process; literary texts can be repositories for such cultural memories, but can also reflect critically on flood events and the discourse on risk and disaster. In the introduction to his edited volume *Criticism, Crisis and Contemporary Narrative*, Paul Crosthwaite has pointed out the relationship between literary criticism and crisis, noting that "the interface of narrative dramatizations of life under conditions of global uncertainty and instability and the practices of literary and cultural criticism yield forms of knowledge—experiential, affective, libidinal—not readily accessed via the objects of study or methodologies of other disciplines" (2011, 5). The crises invoked in this chapter are both the immediate crises of storm surges and the more abstract crises of collective national identity and understandings of risk, both of which reaffirm certain ideas about the North Sea landscape, its history and its future.

This chapter will explore the crises of living with the North Sea by means of literary texts spanning more than 100 years that show how the history of the hazardous landscape is entangled with narratives of nationhood and national development. Hans Christian Andersen's novel *De to Baronesser* (*The Two Baronesses*) from 1848 advances the then-contested province of Schleswig at the heart of a resurgent Denmark with strong ties across the North Sea; this is compared with Arthur van Schendel's historical novel *De Waterman* (*The Waterman*) from 1933, which depicts the struggles of the Netherlands during the nineteenth century as seen from the crisis of interwar Europe. The final part of the chapter explores literary responses to the catastrophic storm surge of 1953, and contemplates the way that the discourse on this "natural disaster" engages with late-twentieth-century ideas about environmental resilience in the North Sea. (The term natural disaster is rightly criticized by scholars of disaster, as "natural forces become disastrous only within social and cultural frameworks" (Mauelshagen 2009, 42), and for that reason I have avoided using it in this book.)

De to Baronesser (1848)

Hans Christian Andersen (1805–75) is generally remembered for his fairy tales and travelogues rather than his novels, of which he wrote only six during his life. These novels are more clearly rooted in Denmark's geography and society than his cosmopolitan fairy-tale output, and the geographical is an important component of the text itself (Behschnitt 2006, 392). *De to Baronesser* has a complicated plot, involving an orphaned baby girl who is brought up as a ward of four young gentlemen of Danish society, one of whom is himself orphaned and whose upbringing is the charge of his idiosyncratic grandmother. The baby girl, Elisabeth, and the grandmother are the two baronesses of the title, the irony being that both were born into poverty and made baronesses by twists of fate; more importantly, they are united by their strength of character. The story thus deals with lost, mistaken, and created identities, and the way these are formed in Danish society in the nineteenth century.

Denmark during this period was in the process of redefining itself as a nation; the earlier part of the nineteenth century had seen Denmark suffer defeat in the Napoleonic wars, and 1848, the year this novel was first published, saw the outbreak of the First Schleswig War (Three Years' War). Schleswig-Holstein was a contested region throughout the nineteenth century; although Denmark retained control of the area at the end of hostilities in 1851, this proved to be little more than a ceasefire. Following the Second Schleswig War of 1864, the territories fell under German control. Andersen's choice of location was thus a political one, even if, as Behschnitt asserts, his primary motivation was to exploit the exotic landscape for literary purposes (2006, 408).

Denmark's political difficulties in the nineteenth century came on the heels of an "ecological crisis" that reached its apex in the eighteenth century, when the intensive exploitation of the landscape for agriculture, coupled with deforestation to provide wood for shipbuilding, led to severe sandstorms and wind erosion; in other areas, the deforestation caused agricultural land to be waterlogged (Sörlin and Öckerman 1998, 33–7). By the beginning of the nineteenth century, techniques to bind soil and sand were improving resilience but perceptions of landscape instability and change had left their mark. Hans Christian Andersen wrote about the shifting sand dunes in "En Historie fra Klitterne" from 1860 (published in English as "A Story from the Sand Dunes").

There is thus a parallel between the shifting fortunes of Andersen's protagonists, the unstable landscape, and the nation struggling with its recent history and sense of itself on the European stage. The two baronesses and the grandson, Herman, who is seemingly the bastard son of an Italian highwayman rather than the high-born orphan he believes himself to be at the outset, are each implicitly trying to make good their history of poverty and low birth. The old baroness's successful recreation of herself is dependent on her victory over the baron who oppressed her parents; she avenges cruelty by acts of peculiar kindness to the peasantry around her estate, and ultimately the old baroness is successful in raising herself morally as well as in terms of social mobility. Herman's search for his identity leads him for a long time to reject his Danish birthplace and seek solace abroad. Elisabeth's identity is formed by her foster parents on the bleak Danish hallig (tidal island) of Oland, in Schleswig—defiantly Danish at the time the novel was written—and it is her pious and innocent character, shaped by this wild natural outpost of Denmark, that make her the perfect Danish bride for Herman. At the end of the novel he returns to his own Danish homeland, marries Elisabeth, and takes up his pre-ordained role managing his grandmother's and his own estates.

The action moves from the island of Langeland (where the weather drives Herman and his friends to take an unplanned detour) to Funen (Fyn) to Copenhagen, and thence to the halligs, and then back to Copenhagen and finally Funen again; Oland, which is the most marginal space geographically, is thus structurally at the heart of the novel. Wolfgang Behschnitt has noted a correlation between the narrative structure of the novel and the movement through the geographical space, a symmetrical and cyclical journey in a space that extends from the Danish capital city to the outmost periphery of the nation, echoing Elisabeth's tempestuous journey to adulthood (Behschnitt 2006, 414). It is Oland where Elisabeth is raised from a small child to a young woman, before she returns to Denmark proper to marry Herman and fulfil the will of the old baroness. This flat and windswept hallig is shown as being pivotal to Elisabeth's maturing into a pious and upstanding adult. Oland is defined by its distance from Danish society and its exposure to the forces of nature. Whereas Funen is shown in the text to be the cradle of Old Norse culture in Denmark (see Andersen 1866, 26–7; 1870, 24–5),[1] the North Sea coast of Jutland is the cradle of geopolitical allegiances, those of the North Sea countries, in particular Scotland. When Elisabeth accidentally stows away in a carriage on her way from mainland Denmark (Jutland) to Föhr, she meets innkeepers on the North Sea coast who have gained a new piece of land overnight:

> En faaer Godt og En faaer Ondt af Natten! I fik inat en Unge og jeg fik forleden Nat et Stykke høit Marskland, der drev op paa min Grund.
>
> (One had got a good bargain, and another a bad one at night! You got a young one last night, and I got the other night a piece of high Marskland, that was driven up on my ground.)
>
> (110; 104)

The innkeeper continues:

> "I maa see mit Stykke høie Marskland; det kommer lige fra Skotland eller Island."
>
> Den Vinding af Land, Havet havde givet ham, var hans Tanke og Stolthed. "Das hohe Moor," som det kaldes, to svømmende Øer, var lagte til hans Grund. Efter Folketroen kommer dette fra Island eller fra Skotlands Kyst, men det er simplere at forklare, og dertil det eneste Rigtige, at det er Dele af det undergaaede Friisland, som af Havet selv er løftet fra dets Bund, hvor det laa, og som komme drivende op og blive siddende paa Sandbankerne.

("You must see my piece of high Marskland; it comes straight from Scotland, or Iceland!"

This piece of land, which the sea had given him, was his continual thought and pride. *Das hohe Moor*, as it is called—two swimming islands were added to his land. According to the belief of the people there, they came from the coast of Iceland, or from Scotland; but it is more easily explained, and also more correctly, by supposing what is the only just opinion, that these pieces of land are part of Friesland, which has sunk, and, lifted up by the sea itself from its bottom, where it lay, drives about and settles on the sand banks.)

(111; 105)

This is the first mention of a folk belief in the contiguousness of culture and indeed of actual land between Scotland (and Iceland) and Denmark. This is picked up again a little later, when the protagonists on Oland read the novels of Walter Scott:

> Det Læste blev overført paa hvad de selv kjendte og havde oplevet; det Beslægtede mellem Skotterne og Friserne laa saa nær, ikke blot i Sproget, men i gamle Skikke og Brug.
>
> (The descriptions were transferred to what they themselves knew and had experienced: the connection between the Scotch and the Frisians was so close, not alone in the language, but in old habits and customs.)
>
> (126; 120)

And a later visitor to Föhr from Scotland finds himself on familiar ground:

> Han havde i ung Alder bereist hele Europa, men i det sidste Aar var det især Tydskland, Holland og det skandinaviske Norden, som havde hans hele Interesse; det Beslægtede i disse Nationer, Familiebaandet mellem dem, var hans Studium.
>
> (He had in his youth travelled through the whole of Europe, but of late years it was Germany, Holland and the Scandinavian north that had especially interested him; the relationship between these nations and their family ties were his study.)
>
> (137; 131)

The sense of community that Hans Christian Andersen creates amongst the nations that border the North Sea belie the recent history of northern Frisia as German, rather than Danish territory, and present the history of the halligs as part of an ancient North Sea culture in which Denmark has as much stake as any other coastal nation. The two ways in which the cultural alliance across the North Sea is articulated—textually through the novels of Scott, and physically through the floating island—underline the reciprocity of landscape, something constituted by both environmental and imaginative processes.

The nature theology that is woven through Elisabeth's experience of Oland also shows the way in which landscape can be read as text, and vice versa. Elisabeth is brought up from the age of five by her foster parents: Moritz, a pastor, whose young bride died before she could accompany him to his parish on the hallig, and Moritz's widowed sister Hedevig. Moritz supplies the perfect example of the nature theology that had become popular in the eighteenth and early nineteenth centuries, and the hallig of Oland is the perfect landscape for his understanding of "de to Bøger, Natur og Bibel" (135)—the two great books of Scripture and creation—the contemplation of which bring him close to the sublime:

> Ved denne Leven ind i alle Bøgers Bøger, ved fra disse at kunne læse høit for sin Kreds og Menighed, saaledes som Moritz af sit Hjertes Trang maatte, var det ikke kun fra Kirkens Prædikestol, han pegede hen paa Gud, men fra den lave Strandbred, fra Nabo-Øernes Dyner; man vilde i enhver Tidsalder have forstaaet ham paa disse Steder. Intet betragtede han som Tilfældigt, men som et Led i den store Kjæde … Han lærte dem, hvorledes det enkelte Straahalm, som Vinden kaster om paa Kysten og fæster der, maaskee om Aarhundreder er til Frelse for Landet i den truende Stormflod. Det ubetydelige Straa samler om sig det løse Sand, det voxer og voxer, og tilsidst staaer det som en beskyttende Sanddyne. "Dog, stol ikke paa den," sagde han da, "see i den Magtens og Storhedens Fald! Intet i denne Verden er varigt. Hist ude, hvor Havet bryder sig mod Revlerne, stod for faa Aar tilbage et mægtigt Sandbjerg; da kom Kaninerne, de gravede deri deres Huler, Vinden blæste ind i disse, borede ved Nat og Dag, Søen slog mod den løse Grund, og Sandbjerget sank, rullede ind under Bølgerne."
>
> (This intimacy with the Book of books, and the opportunity of reading aloud therefrom to his private circle and his congregation, just as Moritz's heart dictated, led him, not alone from the pulpit to point to God, but from the low shores of the strand, from the dikes of the neighbour islands—and men in all ages have understood such significant lessons. He regarded nothing as incidental, but, on the contrary, as a link in the great chain.… He taught them how a single weed which the wind drives up on the coast, and which takes root there, may perhaps in centuries become a safeguard against the encroachments of the sea. The insignificant weed gathers the loose sand around it; it grows and grows, and at last stands like a protecting sand-dike. Yet do not depend on such!—in them behold the fall of power and greatness!—nothing in the world is durable. Yonder, where the sea breaks against the shoals, there stood, not many years ago, a mighty sand bank; the rabbits came there and dug their holes, the wind blew in there, and forced its onward way incessantly; the sea struck against the loose ground and the sand bank fell and was engulphed [*sic*] by the ocean.)
>
> (136; 129)

The landscape of the hallig, and by extension of the windswept coasts of the other nations bordering the North Sea, is one united by its particular impermanence, as this passage indicates. This passage also illustrates beautifully how the literary and real landscapes are co-constructed; Moritz, a literary creation, uses the landscape of the hallig as a means of interpreting scripture. The fact that the halligen are in a permanent state of flux is a feature of life there and familiar to all inhabitants. The Commander and Madam Levsen, inhabitants of Oland, have a pulpit in their house beneath which the Commander was baptised and the pair were wed, a relic from the old church in Rantum on Sylt.

> "Tre Gange blev Kirken flyttet ind i Landet," sagde Commandoren, "tre Gange blev Byen flyttet ind, og det var saa stor en By, at den engang i Storm mistede flere end to hundrede Baade; Sandstormen driver indad og Søen løber bag efter og tager bort; den vil tage os Allesammen, naar vi ere i vor Grav."
>
> ("Three times was the church moved further into the land," said the Commander, "and three times was the town moved further in, and it was so large a town that it once lost two hundred boats in a storm. The sand-storm drives inwards, and the sea follows it and carries away all in its course; it will take us altogether, when we are in our graves.")
>
> (107; 100)

This is something Elisabeth discovers about the hallig soon after she arrives, when the housemaid Keike is charged with showing her round. Keike takes her to the old and new graveyards on Oland, the first of which is in the process of being washed away by the sea. Keike gathers some loose bones from the seaward edge of the graveyard in her apron in order to transport them to the new graveyard at the other end of the island, "saa vare de dog i Ro saalænge, til Havet engang naaede derop" ("so that they would at least rest in peace until the sea reached so far") (114; 108).

Keike has experienced a storm surge, spending two nights perched on a house top awaiting rescue: she also knows the legend of Rungholt, the Frisian settlement finally destroyed in storm surges in the second "Grote Mandränke" in the sixteenth century, and links this to the modern-day inhabitants of the halligs.

> Hele Rungholdt og syv Kirkebyer sank, og kun Præsten, hans Pige og to Jomfruer, som havde været i Kirke, undslap. "Slægten lever endnu," tilføiede Keike, "det er Bakke Boyesens. Og vist er det, at engang før Dommedag vil Rungholdt hæve sig med alle dem deri op af Havet. Ved klart Vande øiner man endnu Huse, Kirketaarne og Møller; jeg har ikke selv seet det, for jeg lukkede Øinene, men det ringede for mine Ører, det var Klokkerne dernede, som gik."

> (All Rungholt and seven church-villages sank; the parson, his servant-maid, and two young girls who had been in the church, alone escaped. "That race still lives," added Keike, "they are called Boyesen; and it is sure, that before the day of judgement, Rungholt will rise again from the sea with all that were in it. When the water is clear, one can yet see houses, church-towers, and mills; I have not seen them myself, for I shut my eyes, but there was such a ringing in my ears,—it was the church bells that rung!")
>
> (116; 110)

The uncovered graves and imagination of underwater church towers are classic *memento mori* motifs, anchored in the pietistic spirituality of the region. They imbue the landscape, in Hans Christian Andersen's portrayal, with its readability as a cipher for God's creation. Another stark image of the transience of human existence is provided in the novel by the perspective provided by the telescope owned by the Commander; this instrument, a technical marvel that was part of the humanist advance in understanding the natural world, provides a distanced, bird's-eye perspective from which to view the markers of impermanence in the dynamic landscape:

> I Storm og Stranding kommer den [Kikkerten] frem og fortæller om Død eller Redning; ved langvarig Østenvind, naar Havet er blæst bort milevidt i Ebben, og Rudera af undergaaede Byer, Skrogene af strandede Skibe, Hvalfiskeskeletter, Havnepladse, Kirkemure og Begravelsesstene rage spøgelseagtigt op af det hvide Sand…
>
> (It [the telescope] is drawn forth in storm and shipwreck, and tells of death or rescue; during a long and continued east wind, when the sea is blown back for miles into an ebb, and the remains of sunken towns, the hulls of stranded vessels, and the skeletons of whales, harbors, church walls and tomb-stones protrude like ghosts from out the red sand…)
>
> (108; 102)

The defining features of this landscape are its constant reshaping, both by slow processes (erosion, siltation) and sudden events, such as sandstorms, the appearance of floating islands, and storm surges. The portrait of the landscape is provided in a collage of intertexts that underline the interrelationship between text and place. Hans Christian Andersen's sense of a common heritage in the form of a North Sea is drawn from nature theology—the Bible—and literature, e.g., through the novels of Walter Scott, as well as from local folklore. Andersen's inhabitants on Oland have their own version of the legend of the great flood that inundated Doggerland and thus divided the countries of the North Sea that had once been part of the same land mass.

> "Men Landene have jo ogsaa engang hængt sammen," sagde Madam Levsen; "idetmindste har England hængt fast ved Slesvig og Jylland. Det er et

> gammelt Sagn, at en heel Række Høie bandt England til vort Land; og den engelske Dronning skulde giftes med den danske Konge, men han sveg hende; da lod hun i syv Aar, med syv hundrede Mand, grave gjennem de grønne Høie, der holdt Havet paa begge Sider, og saa gik Havet sammen, slugte alle Øer og svømmede over hele Friisland, der blev til det Øland, det nu er."
>
> ("But these lands, you know, were once united," said Madam Leyson; "at least, England joined to Sleswick and Jutland. It is an old *saga*, that a whole chain of mountains united England with our land; and the English queen was to have been married to the Danish king, but he deceived her. Then, in her vexation, she ordered seven hundred men to dig through the green hills that restrained the sea on either side, and the seas rushed through and joined, swallowed up all the islands, and overflowed the whole of Friesland, which became the island it now is.")
>
> (127; 121)

This type of legend has been explored by the historian Bernd Rieken, who shows its importance for the North Sea communities' sense of themselves and where they come from (Rieken 2005, 189). Storm surges and floods in particular are defining events in the history of the landscape, and in the lives of its inhabitants: such legends also indicate the importance of storm surges in shaping not just the land, but the historical and literary imagination of a particular society. The legend repeated by Hans Christian Andersen is the mythical story of a landscape shaped by political expediency, rather than divine or natural forces, and it finds an echo in the political motivation of Denmark's most celebrated writer setting his novel in the contested region of Schleswig.

Elisabeth has her own experience of the sea and its capacity to reshape human lives when she and her companion Elimar are caught out by fog during a visit to the neighbouring island of Amrum and almost drowned by the incoming tide. Elisabeth's near-drowning is, in a sense, a baptism into the nature theology that is to shape her character henceforth, under Hedevig's watchful eye;

> Det var Hedevigs Glæde at aabne og skjærpe hendes Øie for naturen; den og Bibelen vare de to Bøger, som altid laae opslagne, hiin uden om, denne inde i huset; begge bleve læste, begge lastede en Afglands i dette Hjem og gjennemtrængte dem, som levede der.
>
> (It was Hedevig's delight to direct her intelligent observations to nature; that and the Bible were the two books that always lay opened, the first around them, and the last within the house. Both were read, both reflected a lustre on that home, and penetrated those who dwelt there.)
>
> (135; 128)

Elisabeth's love for her simple home in Northern Friesland is what sustains her later in the novel, when she is exposed to the distractions and complexities of life in Copenhagen. When Herman returns from his long sojourn abroad, he has rediscovered his loyalty to his Danish homeland; in his own words:

> "ved at reise i fremmede Lande, see hvad Verden kalder det Største og Bedste, kommer man ud af falske Illusioner om noget aldeles Fuldkomment; man faaer Øinene ret op, og det er just ude at Kjærligheden til vort Fædreland bliver stærkere"
>
> ("by travelling in foreign lands, and seeing what the world calls the greatest and best, a man gets rid of the illusion of perfectibility; his eyes are thoroughly opened, and it is then that his love of his country becomes stronger.")
>
> (222; 211)

Denmark has its beechwoods, and it has the sea, and as Councillor Heimeran adds:

> "Ja vort lille Land er et velsignet Land! man veed det ikke selv! Der har nu ogsaa Elisabeth fortalt mig om Oland og Amrom, om disse Øer under det Slesvigske, og var det ikke, som om jeg hørte om Lande i andre Verdens-Dele? thi hun fortæller ogsaa godt."
>
> ("Yes, our little land is a blessed land; one does not know it oneself. Elisabeth has told me about Oland and Amrom, about those islands near Sleswick, and it was as if I heard her tell about lands in other parts of the worlds; for she can tell a good story very well.")
>
> (224; 212)

Innocent and unspoilt Elisabeth is elided here with the windswept and ephemeral margin of Denmark, "vildt og nøgent" ("wild and bare") (224; 212), which is just as wild and beautiful as anything else the world has to offer. Her own good nature is closely tied to the nature of the halligs, as she muses:

> "man tænkte der mere paa vor Herre! jeg troer, Menneskene her i Kjøbenhavn ere vist i Hjertet ligesaa gudsfrygtige, som de derovre, men de faae ikke her Tid at øve det … Gud kom der mere ind i Ord og Tale; her hører jeg ham kun nævne om Søndagen i Kirken."
>
> ("one thought more there of the Almighty! I believe that persons here in Copenhagen are as religious at heart as those over there; but they have not time here to practice it [*sic*] … God came there more into the words and speech. Here I only hear him named on Sundays in the church.")
>
> (226; 214)

Elisabeth is a cipher for the soul of Denmark, which belongs to the transient and contested landscape of its margins more than in the urban environs of

Copenhagen. Herman's return to this reimagined homeland allows him to cast off the shame of his illegitimacy and remake himself in a new image.

Hans Christian Andersen's novel shows how landscapes are constructed in the literary imagination and figure in political and cultural discourse. Denmark's southwestern edge, physically shifting and politically contested, is made and remade as a textual space. The construction of the literary landscape serves to consolidate the understanding of this space as Danish, and northern Frisia as ethnically and culturally allied to the North Sea rather than to Germany. The hardships of life on the North Sea lowlands are recast as virtues that build character and nationalist loyalty in the Danish protagonists. When framed by the honest and homespun wisdom of these "good" Danish protagonists, the *memento mori* symbols of human transience—the pulpit, the graves—become romantic motifs that confer aesthetic, spiritual, and patriotic value on the practices of living on the vulnerable coast.

De Waterman (1933)

Arthur van Schendel's historical novel *De Waterman* also concerns a watery and unstable landscape and religiously inflected risk culture that is full of hardship and impermanence, but rather than illustrating virtue or solace, the struggle to manage the water and reduce the risk of floods is seen as symptomatic of a culture that is inflexible and unwilling to change. Arthur van Schendel (1874–1946) was concerned with the state of his home nation of the Netherlands; he wrote *De Waterman* in 1933, in a period of his creative life when he turned from neo-romantic novels set in southern Europe to more sober and contained works set in Holland. This work is lauded for its "powerful evocation of the Dutch landscape, the rivers and polders and sombre skies" (Meijer 1971, 287) and, as the title suggests, its main character is linked to the water that is a dynamic actor in the history of this nation. Unlike the positive, legitimizing imagination of the halligs as Danish in *De to Baronesser*, the waterscapes in *De Waterman* are imagined as showing a nation constrained by hidebound tradition and unable to fight back against the power of water. The Netherlands' constant struggle with flooding is a metaphorical expression of a failure to embrace enlightened modernity and emancipation and the country's thrall to foreign occupiers.

The novel is set in the early nineteenth century, when the Kingdom of the Netherlands, having regained its sovereignty following the Napoleonic Wars, is suffering from economic and social stagnation and a widespread religious dissatisfaction on the part of both Catholic and Protestant communities (Wielenga 2012, 242–331). The "waterman" of the title is Maarten Rossart, twice outcast—first from his family for questioning the authority of both his father and his starkly Calvinist interpretation of God and Scripture, and second for his decision to pursue and marry a Catholic woman, despite his Reformed beliefs. Maarten is characterized from childhood by an instinctive understanding of the ways of water: he grows up in Gorkum (Gorinchem), where the many rivers and canals of the Rhine–Meuse–Scheldt delta cut through a sodden and hazardous landscape. He learns first to fear the water, when it conceals the body of a French customs official he sees murdered while he is out alone on the dyke. Subsequently, building on Maarten's deep unease about the power of the water, a terrible winter flood arrives. After a hard frost that causes the rivers to freeze over, fear of the coming thaw and the flooding it will bring builds. The slow approach of the disaster is recognized, but the inhabitants of the settlements are helpless in the face of the coming flood:

> de tijding werd van mond tot mond herhaald: bij Nijmegen was het aan het kruien, het water stond er twintig voet, bij Oosterhout scheen het niet te houden, bij Druten steigerden de schollen.
>
> (the news was repeated from mouth to mouth: at Nijmegen it was full of drift-ice, the water was twenty feet high; at Oosterhout it looked as if it would no longer hold; at Druten the ice-floes were rearing up.)
>
> (van Schendel 1940, 28; 1963, 24)[2]

The situation continues to worsen:

> Het regenwater kletterde uit de goot. Het kon wel erger worden dan voor drie jaren, zeide zij, hem wijzend door het dakvenster uit te kijken. Recht voor hem staken de punten van ijsschotsen boven den dijk.
> Het was een Zondag van angsten in het dorp ... mannen met spaden en planken liepen haastig naar de plek waar al van den avond te voren gewerkt was. Daar waren zij bezig een kisting te maken aan de binnenzijde, want buiten viel niets meer te doen, het ijs stapelde zich op, gedurig voortgestuwd, en viel soms in groote schollen zwaar over den dijk. Ginder op den stroom was het breken en barsten te hooren...
>
> ...Hij worstelde, hij wilde er uit naar het gillen van zijn moeder en zijn zusje buiten. Het water steeg snel in de kamer, ook de grootmoeder greep hem, hij werd naar boven getrokken.... Een berg van ijsschotsen stortte tegen het huis dat kraakte en schudde dat de dakbinten braken, de pannen vielen. De klok luidde gedurig ... hij werd langs de ladder naar beneden gelaten en viel in de schuit. Hij zag ijs overal, opgestapeld, kantelend, voortgestuwd, ginds den toren en daken, maar nergens land, nergens den dijk. Hij stond recht op en hij zocht, er was niets dan water en ijs, ginds huilende menschen en mannen die riepen, en overal bulkten de koeien in nood.
>
> (The rain-water pelted from the gutter. "It might get worse than three years ago," she said, taking him to look out of the attic window. Right in front of him the sharp points of the ice-floes stuck out above the dike.
>
> It was a Sunday filled with fears.... men with spades and boards hastily walked towards the spot where ever since the previous evening work had

> been in progress. They were making a coffer-dam there on the inside, for outside nothing could be done any more. The ice was piling up, continuously being pushed along, and sometimes it fell heavily across the dike in huge floes. Beyond, on the river, the breaking and bursting could be heard.
>
> ... He struggled, anxious to get outside to the screaming of his mother and little sister. The water mounted quickly in the room. His grandmother held at him too. He was pulled upstairs ... A mountain of ice-floes tumbled against the house so that it creaked and shook, its roof-beams breaking and the tiles falling. The bells tolled unceasingly.... He was lifted down along the ladder and fell into the boat. He saw ice everywhere, piled up, toppling over, pushed along, and beyond that the tower and roofs, but nowhere land, nowhere the dike. He stood up and searched about: there was nothing but water and ice, people crying and men shouting, and everywhere a lowing of cows in distress.)
>
> (31, 33; 26, 27)

The thaw and heavy rains cause the rivers to burst their banks and the dykes to fail, and Maarten's mother and sister are drowned. In the novel, the agency of water functions as the executor of these characters' fate—one which Maarten himself ultimately cannot escape. His life is intrinsically bound up with the waterways of the southern lowlands. "De Heer strafte in ons land harder met het water dan met het vuur" ("In our country, the Lord punishes more severely with water than with fire") (52; 39), Maarten is told by his father.

As he grows up, Maarten slowly begins to understand the way that water works, and this demystifies it for him; and analogue to this new, modern understanding, the idea of a jealous God poised to punish humanity for their endless sins starts to seem hollow. He works on the dykes with the other men and gains a reputation for his affinity with the water. "Maarten verstond het water, het was hem aangeboren" ("Maarten understood the water, with him it was something innate") (55; 41). Maarten is the only one who will remove his shoes to wade into the waterways and check the dams, allowing him to feel out the strength of the currents and the constituency of the mud, but this also earns him the scorn of his detractors (54, 58; 40, 43). Maarten is an autodidact and empiricist, an Enlightenment figure cut off from the Enlightenment. His independence of mind and empiricism, the values that allow him to work out ways of managing water, bring him into conflict with the community's Calvinist beliefs. Maarten questions the authority of both the Church and the accepted expertise on dyke management:

> Ik ben een gebrekkig mensch, ik kan niet alles verantwoorden wat ik doe. Ik ken de Schrift niet goed.... Als ik kousen en broek uittrek om mijn werk te doen zooals ik dat het beste kan, en het is alweer een zonde, mij goed, dat zal ik dan ook wel met God verrekenen. Dominee moest toch weten dat niet alle menschen kwaad in den zin hebben.
>
> (57)

> (I am an imperfect human being. I cannot justify all I do. I don't know the Scriptures very well.... If I take off my stockings and trousers in order to do my work as best I can, and it is a sin again, then I don't care.—That, too, I shall have to settle with God. Parson surely must know that not all people have evil intentions).
>
> (57; 43)

The many references to Scripture, the text of which is so central to Maarten's Calvinist community, point as in Hans Christian Andersen's novel to the interplay between landscape and text. The lowlands are permeated and sodden not just with water but with the Word, and the Bible is a source of guidance in life in Dutch society: the landscape is made and unmade with the symbolism of a jealous God who judges and punishes. Both *De to Baronesser* and *De Waterman* show how one text in particular produces and reflects on human perceptions of landscape, but also how varied these inflections can be and how the interaction of text and landscape is entangled in the cultural and political climate in which they are both written and read.

Maarten cannot settle into life on the land; he buys a barge, and lives on the water, taking his wife with him. The only time he moves back onto land is to join a brotherhood (one closely modelled on the nineteenth-century Brotherhood of the Prophets of a New Light; an indirect reference to the Enlightenment that Maarten is searching for; Clegg 1963, 6); for a while, their utopian community seems to offer Maarten a new home, but it disintegrates into squabbles over money and doctrine in the face of persecution by the authorities, and Maarten takes to the water once more. This short attempt by a community to raise itself from the mud and reinvent itself as the bright future of the nation ends in failure, and Maarten is not able to emancipate himself or anyone else from the struggle with the water. In the end, it is the water that takes the life of Maarten's only child, who drowns while playing with other children, and Maarten's prophecy that he is destined for a watery grave is also fulfilled at the end of the novel, when, as an old man, he jumps into the water to rescue a dog and slips away to his death.

The environmental historian Petra van Dam refers to the history of human settlement in the Netherlands as "amphibious culture" (2016); this term covers infrastructure and mobility patterns in a watery landscape, but also strategies to cope in the short term with smaller- and larger-scale storm surges—means of evacuating humans and cattle, refuges, systems of providing labour to check and repair dykes. *De Waterman* is a novel that shows the workings of an amphibious culture at a moment of political crisis, just before the transition to a new way of living. Van Dam notes that Dutch amphibious culture started to disappear in the nineteenth century, with the advent of new technologies and construction materials (van Dam 2016, 89), a process that was dramatically accelerated in the latter part of the twentieth century; along with these labour-saving inventions, many humans were emancipated from poverty and regular flooding. But the agency of water in *De Waterman* is dark, cold, and destructive, taking lives without reason and turning humans against one another. Water appears as antagonist, countering Maarten's attempts to emancipate and enlighten himself and others. Unlike in *De to Baronesser*, the inhabitants of the Rhine–Meuse–Scheldt delta in van Schendel's portrayal are not edified by requirements of the mutable landscape and regular floods, but morally stunted.

As a historical novel, it also illuminates the preoccupations of the Netherlands and Europe in the 1930s, when it was written; van Schendel's extension of the old amphibious culture as a metaphor for stagnation and conservatism can also be understood as a critique of the stultified politics of old interwar Europe that were failing to overcome the legacy of miserable (and muddy) trench warfare and prevent the descent into old patterns of militarization, mistrust, and bloodshed. The novel serves to show us how van Schendel employed the North Sea lowlands as a metaphor for a morally and politically waterlogged country and gives us a landscape that undermines, rather than undergirds, the nationalist cause.

De Ramp in literature

Less than a century—the passage of some 80-odd years—separates the two novels discussed thus far, but the way in which the landscape's past and future are employed is a useful illustration of changing ideas about nature, disaster, technology, and progress over the course of the nineteenth and twentieth centuries. In Hans Christian Andersen's portrayal of northern Frisia, the shifting sands and storm surges of the past, are the stuff of legend: the flooding of Mesolithic Doggerland appears as myth, but Oland's migrating graveyard and the divine wrath that sank Rungholt are also full of mythical symbolism and a semi-mystical view of human transience on Earth. This mysticism is picked up in the nature theology preached by the pastor Moritz, in which all of Creation has been bestowed with immanent meaning. The nature of the North Sea coasts is Danish; and also mysterious, beautiful, and edifying. Van Schendel's historical novel has by definition the past of the region as its subject: in stark contrast to *De to Baronesser*, the amphibious societies of the lowlands were mired in superstition and misery, and this points to ideas about progress and stasis prevalent at the time of the novel's writing in the 1930s. Van Schendel finds no great symbolic meaning, no romantic imagery, in the breached dykes, but a failure of human culture to enlighten itself. The North Sea lowlands serve in the two novels as sites for the imagination of national failures and successes.

The constructed nature of landscape, and the entanglement of imagination in the way North Sea environments and processes are perceived and understood, is brought sharply into focus in the middle of the twentieth century with the most devastating (at least from a human perspective) storm surge within living memory. The twentieth-century narrative of human progress in a lowland environment is brutally checked in the Netherlands in 1953 by the scale of the flooding and the extent of human suffering. While technological advances and state projects of "internal colonization" (van den Grift 2013) in the twentieth century had suggested increasing liberation from the hazards of the lowland geography, the ground on which the Dutch nation-state was built was, after all, still wet and subsiding. The "Watersnoodramp" was a stark reminder that the hazardous landscape had not been tamed.

In the early hours of 1 February 1953, a spring tide combined with a violent storm drove a far larger than normal volume of water into the North Sea basin, causing sudden and catastrophic flooding on the east coast of England and in Holland and Belgium. More than 300 people died in England and there were further losses in Belgium, but in the Netherlands the death toll was almost 2,000 people, along with the destruction of livestock and terrible damage to buildings and infrastructure. The disaster was exacerbated by the lack of any coherent warning system, and the fact that the high tide came at night, when the radio was not broadcasting and most people were in bed. Moreover, dyke maintenance had been neglected in previous decades due to the German occupation of 1940–44, the bombing of Rotterdam, and the 1944 famine, all of which drastically undermined resilience in the post-war decade (Gerritsen 2005). The Watersnoodramp, often referred to simply as "de ramp," was the worst flooding disaster in the Netherlands in the twentieth century and was the driver behind large-scale hydro-engineering projects such as the Delta Works, and its companion in England, the Thames Barrier (Kelly 2018; Gerritsen 2005, 1284–8).

The historian Selma Leydesdorff, one of the chief proponents of memory theory in relation to the 1953 disaster, notes the impact of this flood on Dutch cultural memory: "Recollections of this flood form part of the mythology of the national identity and memory that has to do with centuries of victory over the water" (2000, 76). She argues that traumatic memories of the flood are suppressed in favour of a narrative of technical lessons and future security. Meindert Schroor also notes that the impression that the event has left on the public consciousness focuses exclusively on the successful battle against the water, with no mention made of failures and mistakes (2011, 159).

But literary engagements with the Watersnoodramp paint a more complex picture of the aftermath of the disaster, interlacing texts with the experience of trauma and showing an ambivalence towards myth as an explanatory device. One of the best-known literary portrayals of this flood event is Jan de Hartog's *The Little Ark*, published in 1953, in which the lives of two adopted children are touched by the kindness of strangers in the immediate aftermath of the flooding. *The Little Ark* is ostensibly more concerned with the heroic story of the children's survival than with the trauma of the disaster—the story of the two children and their rescued animals who survive the floods, first in a bell tower and then in a house-boat, was turned into a film in 1972 and is advertised as a children's adventure film (the trailer for the film version of *The Little Ark* boasts that "what [the two child protagonists] learn about courage and love is a lesson for children everywhere").[3] Nonetheless, on closer inspection, it appears to be a novel written *against* the mythification of flood narratives and towards a more cynical view of disaster stories. *The Little Ark* involves a number of slightly fantastical adventures for the two children, and in many ways it conforms to the conventions of children's writing, but it is, as Ad Zuiderent argues, more than a children's book (2003a, 12). The novel contains moments of darkness (the discovery by the children of the floating corpse of their disliked adoptive mother, Mrs Grijpsma, who was drowned trying to reach them) and some wry statements about the nature of unity. In a moment of post-disaster ecumenical compassion, their adoptive father, a strict Calvinist minister, works together with a Roman Catholic priest to provide burials for the many dead. Afterwards, Mr Grijpsma talks to the children:

> "But, you see, here in this broken village, with all those dead people and dead animals around us, there is a wonderful feeling just now of all mankind helping, and, well, thinking of us with love … Father Ambrosius found a note in French.… We kept that note, for to us it expressed the heart of the miracle. It said, '*Notre race c'est l'humanité, notre patrie le monde.*'"
> "What does that mean?" Adinda asked.
> "Our race is mankind, and our fatherland is the world."
> "Your race is white," Adinda said, "and your fatherland is Holland."
> (de Hartog 1953a, 165)[4]

The uncomfortable character of Adinda, with her non-white skin and troubled memories of other places, sees through the myth-making potential of the disaster; and, in fact, the entire novel turns on the subversion of well-worn myths. The animals that survive the flood, unlike the biblical residents of Noah's Ark, are a motley collection of creatures, and there is just one of each of them; the ecumenical truce between the two clerics is abruptly ended when a deranged farmer kills Father Ambrosius with a pitchfork; and, although the children's village of Nieuwerland is being rebuilt from the rubble and mudflats, there is no dove of peace in sight, but rather "a fat, dirty pig, snorting and rolling in a pig's paradise" (209), and the bad-tempered and featherless cockerel, Prince, who has the last line in the novel (213).

The Little Ark has echoes of the critique of religion found in *De Waterman*; biblical Christianity, while ubiquitous and not ill-intended, is shown to be on the defensive, with the clear-sighted orphans Jan and Adinda impassive to its comforts. The mangled scripture quoted by their kindly father is a further sign that mythical-spiritual explanations for the flooding are unfit for purpose. Mr Grijpsma has trouble finding a suitable Bible passage for the funeral of Father Ambrosius, and has to mix and match (199); when he gives the address, he reads it "very badly" (203). The history and future of the natural environment plays little part in this novel, increasing the sense of the flood as an arbitrary event, not part of any longer collective memory of repeated storm surges and breached dykes. The choice to narrate the novel from the perspective of two young children foregrounds the unexpectedness of the flood and the confusion of its aftermath without raising the issue of flood preparedness or the causes of the disaster.

Some of the other immediate responses to the flood are collected in Zuiderent's anthology of writing on 1953 that was published to commemorate the

30th anniversary of "de ramp" (2003b).[5] Much of the writing from this anthology (the first of two) was composed directly after the disaster; one such is the sonnet by Rosa Knorringa, with its final couplet a concise summary of this bewilderment of the poetic subject, a rhetorical appeal to make sense of the disaster, "Hoe moeten wij ooit deze ongerijmdheid boeten/Dat stenen klinken onder onze voeten?" ("What are we to make of this absurdity/that the stones [here] ring under our feet?") (1983, 87). Knorringa's poem is also noteworthy for its setting: the stones that ring under the urban subject's feet are the urban pavements; the landscape of the coasts is both distant and unfamiliar. A similar sense of powerlessness and bewilderment can be found in other texts from the same anthology, such as A. Korthals Altes's poem "Bergingswerk" ("Salvaging"; Zuiderent, 2003b, 10) and Okke Jager's bitter verse "De kindermoord van Zeeland" ("The Massacre of the Innocents in Zeeland"; Zuiderent, 2003b, 83; the latter poem draws on the imagery of the infants slaughtered by King Herod in the Gospel of Matthew, hence my translation of the title). The impression throughout these poems is that the storm surge appears to have come from nowhere to wreak its havoc. There is no one to blame, and no solace to be found.

A much later and more reflected response to the disaster is the 2005 novel *De verdronkene* by Margriet de Moor. *De verdronkene* ("The Drowned") also explores the undigested grief of 1953 through the lives of two sisters caught up in the tragedy of the Watersnoodramp (de Moor 2007). It picks up on the trauma evident in the poems collected by Ad Zuiderent, in which grief and bewilderment are the dominant emotions. The novel alternates between the narrative of Lidy, who drowns in 1953, and that of her sister Armanda, who lives a long life; while Armanda's narrative speeds up as she ages, Lidy's slows down to recount the final hours and then minutes of her life before she finally succumbs to the waters. The sudden and unpredictable agency of the water is underlined by the plot in which the two sisters exchange places for no real reason, with Lidy taking up an engagement in Zeeland in place of Armanda, who stays behind in Rotterdam; just as the floodwaters seem to be an arbitrary, external force, so too is Lidy's death seen as an arbitrary, unpredicted event. The perspective is subjective and entirely individual—Armanda's guilt and grief are compounded by the choices she made that inadvertently led to Lidy's death.

Spiritual comfort was scanty in *The Little Ark*; in *De verdronkene* (the French translation of which, by Danielle Losman, was published with the title *Une catastrophe naturelle*)[6] there is no sense at all of any mystical meaning in the transience of human life and the sudden and brute power of the natural world. But the most notable trend across all of these literary portrayals of the 1953 flood is the absence of the environment and the history of the human habitation of the lowlands. The storm surge bursts into human narratives without warning and retreats again, leaving behind grief and disillusionment—but no engagement with the landscape or wider exploration of the human habitation of the vulnerable edges of the North Sea. Selma Leydesdorff notes that:

> How one feels about immunity from natural disaster is relative. Feelings of safety are culturally determined, and our sense of danger varies over time. For example, what is happening to the natural environment was just as menacing twenty years ago as it is today, but now the public is more aware of it. The same is true of crime. And yet the Dutch feel safe and protected from the sea, despite simultaneously being aware of the rising sea level and the melting ice cap.
>
> (Leydesdorff 2000, 77)

In their recent study of risk perception on the German North Sea coast, Beate Ratter and Cilli Sobiech (2011) found that only a third of the coastal population surveyed saw a risk to the area from storm surges and climate change. However, Greg Bankoff, in his influential book on cultures of disaster, adds:

> While a greater appreciation of the importance of hazard to the construction of culture still gains little acceptance among most western scientific communities, there are already signs of changes to come. As the developed world itself is increasingly beset by the consequences of global warming, climate change and rising sea levels, ideas about hazard and disaster may have to be reconsidered.
>
> (2003, 3)

For evidence of a new understanding of the hazards inherent in living in the lowlands we must look to newer texts; one of these, "The Netherlands Lives with Water" will be explored in Chapter 2. The incorporation of flooded landscapes into recent novels such as Isabel Hoving's *De een na laatste dood van het meisje Capone* and Eva Moraal's *Overstroomd*, both from 2015, point to an increasing (re)incursion of the physical environment into Dutch literature.

Perceptions of the risk culture of the North Sea coast have shifted over time, along with the shorelines. The novels and poetry that are responses to and portrayals of the 1953 flood do not carry the history of the region as defined by constant shifts in the balance between land and water in the way that earlier texts do; the metaphor of the struggle of humans against the power of the tide, whether this is seen as contributing to human virtue or to human recalcitrance, is not present. Instead, the storm flood is seen as something beyond the parameters of human knowledge and history. Through their depiction of human responses to the disaster, though, these Dutch texts are able to question a dominant narrative of heroism and technical progress.

Following the textual imagination across time also calls us to consider technical "progress" in managing the risks of living in the North Sea lowlands. In the *De Waterman*, Maarten's desire to manage the environment better and emancipate his society from the hazards of living with the water is an expression of the "progress" that saw the conquest of nature as the path to human freedom and betterment. Environmental historians have traced this trajectory of taming watery environments to make them safer for humans (Blackbourn 2007;

Rotherham 2013; van Dam 2016). In his "Four Theses" (2009), Dipesh Chakrabarty posits that in pursuing the conquest of nature that brought about the emancipation of large parts of the population in Western Europe—the kind of emancipation Maarten was trying and failing to achieve—humanity unwittingly made itself the geological agent that brought about the Anthropocene. The aim was to change environmental conditions locally, but the outcome has been to change them globally, for the whole of the planet, and for the worse. The Dutch texts we have seen that explore the response to the 1953 flood capture a moment when narratives of "progress" were shattered, or, as Matthew Kelly puts it, paraphrasing Braudel, "the 1953 flood was a moment when the *longue-durée* became conspicuous" (Kelly 2018, 206). The expressions of human confusion and trauma run counter to the dominant narrative that continued to reproduce the idea of progress (e.g., in engineering solutions such as the Delta Works and the Thames Barrier). In their emphasis on bewilderment, these texts also refuse to make sense of the disaster. Religious and humanistic ideals of altruism in times of crisis are rejected by the children in *The Little Ark*; In *De verdronkene*, the death of Lidy in the flood is portrayed as the result of the arbitrariness of both humans and nature. These responses only capture the moment of shattering impact, when the cause, and the full cost, are still unknown.

Notes

1 All page numbers for quotations in Danish from *De to Baronesser* refer to Andersen (1866); all English quotations are from the translation by Charles Beckwith Lohmeyer (Andersen 1870).
2 All page numbers for quotations in Dutch from *De Waterman* refer to van Schendel (1940); all English quotations are from the translation by Neline Clegg (van Schendel 1963).
3 Film trailer, *The Little Ark* (1972), www.youtube.com/watch?v=-FFxOSIfQNc (last accessed on 14 January 2016).
4 The novel was published in both Dutch and English in the same year. Since Jan de Hartog wrote many of his books in English, I have cited from the English version of *The Little Ark* (de Hartog 1953a). The Dutch version is published as *De kleine Ark* (de Hartog 1953a).
5 Editor Ad Zuiderent subsequently brought out a second anthology, 60 years after the disaster, containing more distanced, second-generation responses; see Zuiderent (2003b).
6 This novel was published in German as *Sturmflut* and in English *The Storm*.

References

Andersen, Hans Christian. 1860. "En Historie fra Klitterne." In *Nye Eventyr og Historier af H. C. Andersen. Fjerde Samling*. Copenhagen: Reitzels Forlag. 35–57.
Andersen, Hans Christian. 1866. *De to Baronesser: Original Roman i tre Dele*. Copenhagen: Reitzels Forlag.
Andersen, Hans Christian. 1870. *The Two Baronesses: A Romance*. Translated by Charles Beckwith Lohmeyer. New York: Hurd & Houghton.
Bankoff, Greg. 2003. *Cultures of Disaster: Society and Natural Hazards in the Philippines*. London: Psychology Press.

Bankoff, Greg. 2013. "The 'English Lowlands' and the North Sea Basin System: A History of Shared Risk." *Environment and History* 19 (1): 3–37.
Behschnitt, Wolfgang. 2006. *Wanderungen mit der Wünschelrute: Landesbeschreibende Literatur die vorgestellte Geographie Deutschlands und Dänemarks im 19. Jahrhundert*. Würzburg: Ergon Verlag.
Blackbourn, David. 2007. *The Conquest of Nature: Water, Landscape and the Making of Modern Germany*. London: Pimlico.
Chakrabarty, Dipesh. 2009. "The Climate of History: Four Theses." *Critical Inquiry* 35, no. 2 (Winter): 197–222.
Clegg, Neline C. 1963. "Introduction." In *The Waterman*, by Artur van Schendel, 5–7. Translated by Neline C. Clegg. Leyden: A.W. Sythoff.
Crosthwaite, Paul. 2011. "Introduction." In *Criticism, Crisis and Contemporary Narrative*, edited by Paul Crosthwaite, 1–11. New York: Routledge.
de Hartog, Jan. 1953a. *The Little Ark*. New York: Harper & Brothers.
de Hartog, Jan. 1953b. *De kleine Ark*. Amsterdam: Elsevier.
de Moor, Margriet. 2007. *De verdronkene*. Amsterdam: Contact.
Gerritsen, Herman. 2005. "What Happened in 1953? The Big Flood in the Netherlands in Retrospect." *Philosophical Transactions of the Royal Society* A 363, 1271–91.
Hoving, Isabel. 2015. *De een na laatste dood van het meisje Capone*. Amsterdam: Querido.
Hulme, Mike. 2009. *Why We Disagree About Climate Change: Understanding Controversy, Inaction, and Opportunity*. Cambridge: Cambridge University Press.
Huntingdon, Henry of. 1753. *The Chronicles of Henry of Huntingon*, edited and translated by Thomas Forester. London: Henry G. Bohn.
Kelly, Matthew. 2018. "The Thames Barrier: Climate Change, Shipping, and the Transition to a New Envirotechnical Regime." In *Histories of Technology, the Environment, and Modern Britain*, edited by Jon Agar and Jacob Ward, 206–28. London: UCL Press.
Kempe, Michael. 2007. "'Mind the Next Flood': Memories of Natural Disasters in Northern Germany from the Sixteenth Century to the Present." *The Medieval History Journal* 10 (1–2): 327–54.
Knorringa, Rosa. 1983. "Op droge grond." In *1 februari 1953: Stormramp en watersnood nagewerkt in gedichten, verhalen en toneeltekst*, edited by Ad Zuiderent, 87. Middelburg: Fanoy Boeken.
Leydesdorff, Selma. 2000. "When the Water Comes: Memories of Survival after the 1953 Flood." In *The Roots of Environmental Consciousness: Popular Tradition and Personal Experience*, edited by Stephen Hussey and Paul Thompson, 76–91. London: Routledge.
Mauelshagen, Franz. 2007. "Flood Disasters and Political Culture at the German North Sea Coast: A Long-term Historical Perspective." *Historical Social Research* 32 (3): 133–44.
Mauelshagen, Franz. 2009. "Disaster and Political Culture in Germany Since 1500." In *Natural Disasters, Cultural Responses: A World History*, edited by Christof Mauch and Christian Pfister, 41–75. Lanham, MD: Lexington Books.
Meijer, Reinder P. 1971. *Literature of the Low Countries: A Short History of Dutch Literature in the Netherlands and Belgium*. Assen: van Gorcum.
Moraal, Eva. 2015. *Overstroomd*. Rotterdam: Lemniscaat.
Ratter, Beate and Cilli Sobiech. 2011. "Heimat, Umwelt und Gefahren—Wahrnehmung der Bevölkerung an der deutschen Nordseeküste." In *Küstenmentalität und Klimawandel: Küstenwandel als kulturelle und soziale Herausforderung*, edited by Ludwig Fischer and Karsten Reise, 181–95. Munich: Oekom.
Rieken, Bernd. 2005. *Nordsee als Mordsee: Sturmfluten und ihre Bedeutung für die Mentalitätsgeschichte der Friesen*. Münster: Waxmann.

Rotherham, Ian. 2013. *The Lost Fens: England's Greatest Ecological Disaster*. Stroud: The History Press.
Schroor, Meindert. 2011. "Flutkatastrophen und Küstenschutz in der niederländischen Marschlandschaft." In *Küstenmentalität und Klimawandel: Küstenwandel als kulturelle und soziale Herausforderung*, edited by Ludwig Fischer and Karsten Reise, 153–66. Munich: Oekom.
Sörlin, Sverker and Anders Öckerman. 1998. *Jorden en ö: En global miljöhistoria*. Stockholm: Natur och Kultur.
van Dam, Petra. 2016. "An Amphibious Culture: Coping with Floods in the Netherlands." In *Local Places, Global Processes*, edited by Peter Coates, David Moon, and Paul Warde, 78–93. Oxford: Oxbow Books.
van den Grift, Liesbeth. 2013. "On New Land a New Society: Internal Colonisation in the Netherlands, 1918–1940." *Contemporary European History* 22 (4): 609–26.
van Schendel, Arthur. 1940. *De Waterman*. Rotterdam: Nijgh & Van Ditmar.
van Schendel, Arthur. 1963. *The Waterman*. Translated by Neline C. Clegg. Leyden: A.W. Sythoff.
Westcott, Kathryn. 2011. "Is King Canute Misunderstood?" *BBC New Magazine*, 26 May 2011. www.bbc.com/news/magazine-13524677. (Last accessed 25 May 2018.)
Wielenga, Friso. 2012. *Geschichte der Niederlande*. Stuttgart: Reclam.
Zuiderent, Ad, ed. 1983. *1 februari 1953: Stormramp en watersnood nagewerkt in gedichten, verhalen en toneeltekst*. Middelburg: Fanoy Boeken.
Zuiderent, Ad. 2003a. "Watersnood en literatuur." *Literatuur* 20: 11–13.
Zuiderent, Ad, ed. 2003b. *Na de watersnood: schrijvers en dichters en de ramp van 1953*. Amsterdam: Querido.

2 Conquest and control

Engineering the Anthropocene on the North Sea

Human habitation of the North Sea coasts was, as was explored in the previous chapter, deeply influenced by the power of the sea and the regular flooding events that shaped and reshaped the coastline. Human responses to this landscape have attempted to manage and mitigate the risks of inundation. Medieval and early modern communities have been studied for the cooperative systems they built to share and manage water risks—the amphibious culture referred to by historians such as Petra van Dam—and historians of disaster refer to "cultures of disaster" to describe the cultural processes of learning from and trying to prevent recurring crises arising from the hazardous settlement in littoral areas (van Dam 2016; van Tielhof 2015; Bankoff 2013; Mauelshagen 2007).

The understanding of risk and the way that risk has been shared within a particular community has been a driver of change on the coasts; the building of new dams and dikes, the reclaiming of new polders, the adoption of flood warning systems, and the decision to give up territories in the face of risk have required communication and cooperation within human societies. The methods of engineering coastal areas have become more ambitious in scope with the arrival of new technologies; attitudes towards the relationship between humans and the natural environment have had to negotiate the shifting sands of risk perception, construction and settlement patterns, and ideas about human reason, progress, technology, and risk. The two literary texts examined in detail in this chapter are stories of risk and technology on the North Sea coasts in the context of wider debates about the role of humans in understanding and controlling their environment.

Dykes and mastery

The first example is one of the best-known literary works concerned with the processes of land reclamation on the North Sea coasts during the eighteenth and nineteenth centuries. Theodor Storm's *Der Schimmelreiter* is a classic text of German poetic realism. It was first published shortly before the author's death in 1888 and has been published in English in several different editions,[1] although it has never enjoyed the popularity abroad that it has in the German-speaking world (Arndt 2012; see Mullen 1988 on reception in English). There are three

German film versions of the novella, and it features regularly as a set text for German schoolchildren.

Theodor Storm spent most of his life on the German North Sea coast and this novella has at its heart the human actors, but also the processes of dyke management and land reclamation in the eighteenth century. It is a historical novel, written more than a century after the events it describes, and it follows the protagonist Hauke Haien's efforts to introduce new dykes with less steep seaward sides[2] in North Frisia, and reclaim and secure new land for his community. For much of its history (most notably during the Nazi period), it has been read as a heroic tale of the human conquest of nature. Later Storm scholarship concentrated on the novella within the context of nineteenth-century German literary developments, in particular with regard to its stance on religion and employment of the uncanny in a text concerned with rationalism, but the text has only recently begun to be explored in a broader context of research into the modern-day environmental awareness. There has been some research on Storm's own understanding of contemporary dyke technology (Eversberg 2014, 429–32; Barz 2000); ecocritical scholarship has also been instrumental in placing Storm's novella in an environmental context, for example in Heather Sullivan's essay "Dirt Theory and Material Criticism" (2012, esp. pages 515–31), the chapter "Breaking Waves" in Kate Rigby's book *Dancing with Disaster: Environmental Histories, Narratives, and Ethics for Perilous Times* (2015, 84–111) and the chapter on Storm in the edited collection *Readings in the Anthropocene* (Ritson 2017).

The plot of this novella is embedded in the broader narrative of the eighteenth century in which it is set—what historian Franz Mauelshagen has called the "master narrative" of the transition from pre-modern to modern times, which he describes thus:

> The Enlightenment of the eighteenth century, preceded by the Scientific Revolution, challenged the monopoly of religious interpretations of nature and established scientific rationality, thus displacing religious explanations of natural phenomena with scientific theories and introducing the idea that humans could control nature by means of technology.
>
> (Mauelshagen 2007, 58)

Hauke Haien is a complex figure, but he is aligned with the incoming values of rationality and belief in technical progress; Wolfgang Frühwald (1981, 442) has asserted that even the name Hauke Haien is aligned with intellect and understanding, as Hauke is the Frisian form of the given name Hugo, derived from the Middle High German term for mind and meaning. The narrative starts with Hauke as a young child questioning his father's measuring and reckoning. His father sends him to fetch an edition of Euclid from the attic so that he can learn something for himself. The edition, when Hauke reappears with it, is a Dutch one, and he needs a Dutch grammar book so that he can first try and understand the language, and then the content, but we learn that he is successful in reading and absorbing the large part of the book by the time the winter is over. This little scene right at the beginning positions Hauke Haien firmly in the humanistic tradition of learning. The knowledge he wants to possess is in a book, and thus breaks with the oral tradition that provided much pre-modern continuity; and, moreover, the book is a classical text by a Greek mathematician, an important cornerstone of the rationalism that underpins Enlightenment thought. The fact that the book is written in Dutch is not coincidental either—apart from testing Hauke Haien's great thirst for knowledge, it is a reminder that new dyke technology in northern Germany was primarily introduced from the Netherlands during the period in question (Blackbourn 2007, 28–9).

Hauke Haien is not just a "geborener Rechner" (Frühwald 1981, 446)—a born mathematician—but is also shown to be a natural empiricist, spending hours outdoors watching the waves strike the sea defences:

> Und wenn im Herbst die Fluten höher stiegen und manch ein Mal die Arbeit eingestellt werden musste, dann ging er nicht mit den andern nach Haus, sondern blieb … an der abfallenden Seite des Deiches sitzen und sah stundenlang zu, wie die trüben Nordseewellen immer höher an die Grasnarbe des Deiches hinaufschlugen.… Nach langem Hinstarren nickte er wohl langsam mit dem Köpfe oder zeichnete, ohne aufzusehen, mit der Hand eine weiche Linie in die Luft.…
>
> (Storm 1888, 14)

> (And when the floods rose higher in the autumn and the work often had to be stopped, he did not return home with the others, but stayed there … on the seaward side of the dike for hours at a time, gazing at the North Sea waves as they beat always higher against the grassy seams of the dike.… After staring downward thus for a long time, he nodded his head slowly; or else, without looking up, he raised his hand and drew a faint line in the air, as though he wanted thereby to give the dike a gentler slope.)
>
> (Storm 1964, 189)

He looks to nature and to the formal rules of geometry and mathematics to understand the world around him. Hauke Haien is shown to be fearless of the natural phenomena that he cannot explain, such as the wraiths that appear on the mudflats in certain winter weathers. These take the form of thin, capering people, and Hauke Haien wonders if they are the souls of the drowned, or the terrible Norwegian sea-ghosts he has been told about, who have a handful of seaweed instead of a countenance, but he refuses to allow himself to believe this. He calls out to them "ihr sollt mich nicht vertreiben!" ("I'm going to stand my ground!") and fights the impulse to run away from them, even though he imagines hearing their "Flügelrauschen und hallendes Geschrei" ("rustle of wings and an echoing cry") (21; 193)[3] at his heels. Much later in the novella, in an interruption from the narrator, we hear Hauke Haien explain to his daughter that these wraiths are merely herons and crows that look much bigger in the fog, searching for fish through cracks in the ice. Hauke Haien's empiricism pays off, and he is able to explain strange phenomena in natural terms.

Besides his autodidactic empiricism, Hauke Haien is clearly marked as an odd man out, someone who is beyond the normal bounds of the small coastal society in which he lives. He is always alone; he studies Euclid on his own and sits alone on the dyke. Cultural knowledge or folklore that is kept and passed on within the community has no place in Hauke Haien's pursuit of science and development. The community itself is shown to be religious, and also divided, with animosity towards some who attend the nonconformist conventicle rather than the church, and towards others still holding to "heathen" practices and superstitious beliefs. Irmgard Roebling (2012, 343), in her recent study of Storm's work, identifies the strongly gendered elements of this dichotomy, with the "Männergeburt" ("male birth") of Hauke Haien (as she points out, his mother is never mentioned) aligning rationality, science, and progress with masculinity, and both the superstitious storytelling, and the dangerously untamed nature, seen to be the preserve of women.

But what initially looks like a simple confrontation between modern rationality on the one hand and pre-modern superstition on the other soon starts to become more complex. Hauke Haien's story first becomes entangled with the superstitious narratives that permeate his society when he kills the beloved angora cat of an elderly widow, Trin' Jans, in a fit of rage. This moment of great irrationality, out of keeping with Hauke Haien's usual demeanour, causes Trin' Jans to curse him ("Du sollst verflucht sein!" ("You'll be cursed!") (26; 195). This event sets up a possible counter-narrative, in which Hauke Haien is not a force of empiricism, but the victim of a curse, of supernatural agencies. The supernatural and rational possibilities for interpretation are vying with each other.

Hauke Haien's curse is not the only way in which he is marked out within the community. He is resented for his success and his fortune in marrying the daughter of the local dykemaster, Elke, through whom he inherits both land and the title of dykemaster for himself. His ambitious plans for a new dyke are met with suspicion. "… von Kindesbeinen an habe ich gehört, der Priel sei nicht zu stopfen, und darum dürfe nicht daran gerührt werden" ("Ever since childhood I've been told that the watercourse out there couldn't be tampered with") (108; 232)[4] are Elke's words, but Hauke Haien, rejecting such folk "knowledge," considers this nothing more than an excuse to sanction laziness. Elke continues:

> Als ich Kind war … hörte ich einmal die Knechte darüber reden; sie meinten, wenn ein Damm dort halten solle, müsse was Lebigs da hineingeworfen und mit verdämmt werden; bei einem Deichbau auf der anderen Seite, vor wohl hundert Jahren, sei ein Zigeunerkind verdämmet worden, das sie um schweres Geld der Mutter abgehandelt hätten; jetzt aber würde wohl keine ihr Kind verkaufen!
>
> (When I was little, I heard the hired men talking about it. They said there was only one way to build a dam out there: to bury something alive in it

while it was being built. They said that, about a hundred years ago when the dike was being built on the other side, the people bought a gypsy child from its mother at a high price and buried the child alive in the dam. But today there isn't any woman hereabout who would sell her child.)

(108; 232)

Hauke Haien of course rejects this superstitious idea, and proceeds with the plans for the new dyke. But he continues to feature in the superstitious stories of the local population. He purchases a horse from a mysterious and unpleasant vagabond; meanwhile, the skeleton of a horse that has long been visible on an uninhabited sandbank out at sea is suddenly seen to have disappeared. The rumours that Hauke Haien has entered a pact with the devil, or is riding the devil's horse, are quickly spread.

Moreover, it seems that Hauke Haien is not just rejecting local folklore, but is also on the verge of rejecting the Almighty. When his wife Elke is delirious with puerperal fever and close to death, Hauke is overheard in prayer:

> Herr, mein Gott … Ich weiß ja wohl, Du kannst nicht allezeit, wie Du willst, auch du nicht; Du bist allweise; du mußt nach Deiner Weisheit thun—o, Herr, sprich nur durch einen Hauch zu mir!
>
> (O Lord, O God … I know You can't always do just what You want to do—not even You can do that. You know everything. Well, Lord, speak to me, speak to me! Just a breath!)

(149; 250)

The rejection of God's omnipotence is added to the other rumours surrounding Hauke Haien, and he is branded a "Gottesleugner" ("outright atheist") (151; 251). Hauke Haien's "unconventionally immanental view of the divine" (Rigby 2015, 103) does not question the existence of God, but it rejects, or at least questions, all forms of power that are not part of his rational worldview, and thus he becomes more and more isolated from his community.

Hauke Haien's grand plans are well meant; he sees his new dyke as adding riches to the community and the new, more gently inclined seaward sides of the dyke as providing greater security from storm surges. But his refusal to compromise his own beliefs, or accept the views of others, mean that his plans find little support, and he is forced to use his authority as dykemaster to initiate work on the dyke. Construction proceeds slowly, but the dyke is eventually finished.

Shortly before the fatal storm that ends the novella, Trin' Jans dies; her parting words, addressed to her long-dead husband, are a warning of what is to come: "Hölp mi! Hölp mi! Du bist ja båwen Wåter … Gott gnåd de annern!" ("Help me! Help me! You're above the water now … God have mercy on all others!") (192; 272). Hauke Haien is unsettled despite himself. "Was wollte die alte Hexe? Sind denn die Sterbenden Propheten?—" ("What did the old witch have in mind? Were dying people able to prophesy?") (198; 273). Trin' Jans, who cursed Hauke Haien as a boy, offers a prophecy of disaster at the end of her life. And disaster soon follows. During the storm that drives the seawater up against the dykes, Hauke Haien finds a group of men about to cut through his new dyke and sacrifice the reclaimed land in order to relieve pressure on the old dyke, which has been neglected. He stops them, and the old dyke breaks, flooding the polder. Hauke Haien sees his wife and daughter swept away by the floods, and throws himself and his horse into the waves, his last words echoing the dying words of Trin' Jans, "Herr Gott, nimm mich; verschon' die Andern!" ("Here, God, take me; but let the others alone!") (218; 282). In this, the moment of his death, Hauke Haien seems at last to capitulate to the powers of both Heaven and nature.

It is hard not to conclude that the novella is at least ambivalent about the power of rational action to bring about change. The land Hauke Haien has reclaimed is still there 100 years later, as the narrator tells us in the final lines of the novel: "Am anderen Morgen, beim goldensten Sonnenlichte, das über einer weiten Verwüstung aufgegangen war, ritt ich über den Hauke-Haien-Deich zur Stadt hinunter" ("Next morning, in the golden sunlight that was shining over a landscape of widespread devastation, I rode my horse down to the city by way of the Hauke Haien Dyke") (222; 284). But in terms of Hauke Haien's personal career, given the flood that swept away the old polder and his own wife and child, his rational project seems to have failed.

The ambivalence within the text regarding the forces of human intellect versus the power of the supernatural is the subject of many scholarly works on Storm. Christian Demandt writes:

> dass man "die Geschichte von Hauke Haien" nicht als Gespenstergeschichte begreifen muss ... soll gar nicht bestritten werden. Zugleich aber sträubt sich der Text gegen eine einfach rationalistische Auflösung der Spuk-Elemente.
>
> (that one doesn't *have* to understand the story of Hauke Haien as a ghost story ... is not the issue. But at the same time, the text resists a simple rational explanation of the supernatural elements.)
>
> (Demandt 2010, 198)

Heather I. Sullivan, in her exploration of the novella's materialist implications, sees Hauke Haien's rationalism as employed to tie down and control the superstitious beliefs of the local inhabitants, "grounding" the uncanny narrative "with precise numbers of dirt-cartloads and rationalist explanations by the schoolteacher narrator" (Sullivan 2012, 526). The reader is forced to engage with both the project of modernity and the obstacles that hinder its progress: it seems that a rational approach to overcoming the hazards of the natural environment and furthering human prosperity is entangled, complex, and problematic, and brings new threats with it.

Mauelshagen, in his essay, warns us against making a too simple dichotomy out of the opposition between superstition and ideas of technical progress:

> The coexistence of religious and scientific ideas, of piety and the quest for technological mastery within particular societies—or even within a particular person—can't be understood if their relationship is reduced to opposition and contradiction. This is true for present as well as past societies.
>
> (Mauelshagen 2007, 61)[5]

This is the problem at the heart of the *Schimmelreiter* novella. Hauke Haien's inability to assimilate the religious worldview and superstitious beliefs of those around him into his own narrative of technical progress and land reclamation on the Frisian coasts is what allows him to become so isolated, and ultimately the driven individual who, almost until the end, is convinced that he and he alone is in possession of true knowledge. His monolithic vision has blind spots; he misses the danger of the breach in the old dyke, and thus brings about the downfall of his family and himself. This reading of the *Schimmelreiter* could certainly draw on parallels between this story and later colonial and neo-colonial engineering projects, in which the knowledge and belief systems of local communities were wilfully ignored in the name of progress, and which subsequently failed dramatically: we remember the 2005 flooding of New Orleans following Hurricane Katrina, in which tried-and-tested methods of living with the danger of floods—local evacuation points, raised housing—were forgotten in favour of high concrete levees, with devastating results (Colten 2012, 159–76; Rigby 2015, 88). Storm's work is, from this perspective, a parable against the dominance of narratives of technical progress and scientific rationality, which alone always have their blind spots; new technical knowledge can make progress in one area (the new dyke profile, the new polder), but, through unintended consequences, cause damage elsewhere (the breach in the old dyke and loss of land).

Since *Der Schimmelreiter* was published in 1888, much has happened to dent human belief in technology and engineering as the means to overcome natural hazards and meet human needs. In the light of climate change and a growing awareness of the dynamism of natural processes, coastal engineering around the North Sea has moved from being largely a story of expansion to one of land preservation and even to the concept of "managed retreat" (sometimes called managed realignment), in which dykes are moved further inland and previously arable fields allowed to revert to floodplain. Writing in the *New York Times*, Michael Kimmelman (2013) makes the case that "the Dutch are starting to let the water in. They are contriving to live with nature, rather than fight (what will inevitably be, they have come to realize) a losing battle."

Jim Shepard's short story "The Netherlands Lives with Water" was first published in 2010 and is set in Rotterdam in the years leading up to a cataclysmic storm that threatens to inundate the city. Rotterdam is in a particularly vulnerable spot on the Rhine–Meuse–Scheldt river delta; parts of the city are six metres below sea level. The short story makes reference to many contemporary engineering features now designed to work with tidal flooding, rather than against it: the "Room for the Rivers" programme,[6] amphibious neighbourhoods, traffic tunnels that double as drainage systems. The story is an ironic commentary on its title, "The Netherlands Lives with Water," which is taken from a Dutch public awareness campaign started in 2003 (see Kazmierczak and Carter 2010, 19–25). As in *Der Schimmelreiter*, the protagonist is involved in technical solutions to lowlands water management, and his understanding of science and the potential of technical solutions to "fix" the problems of increasing flooding are put to the test.

"The Netherlands Lives with Water" is set at an unspecified date in the not-too-distant future, a decade or two after the "great flu epidemic of 2015." Climate change is bringing about rapid change and frequent disaster.

> The World Glacier Monitoring Service, saddled with having to release one glum piece of news after another, had just that year reported that the Pyrenees, Africa, and the Rockies were all glacier-free. The Americans had just confirmed the collapse of the West Antarctic ice sheet. Once-in-a-century floods in England were now occurring every two years. Bangladesh was almost entirely a bay and the whole area a war zone because of the displacement issues.
>
> (Shepard 2011, 41)

The narrator of the story is a Dutch water engineer and his wife, Cato, is a public relations professional whose job is to put positive spin on the issue of flood prevention and control that her husband is working on. Cato is a storyteller, who has to narrate what is basically an imminent disaster scenario in optimistic terms. The communication between the narrator and Cato—thus, by extension, between climate science and public narrative—is explored as both the tension and the water levels rise.

The narrator is, like Hauke Haien, an empiricist. His descriptions of his work and of the conditions of climate change that threaten the existence of Rotterdam are precise and full of measurements and technical detail:

> Sea-facing barriers are inspected both by hand and by laser imaging. Smart dikes schedule their own maintenance based on sensors that detect seepage or changes in pressure and stability. Satellites track ocean currents and water-mass volumes.
>
> (52)

His world is a world of shared knowledge and technical solutions. Rotterdam is constantly placed in context, as the place in Europe where the "hammer" of climate change is hitting first (44); it is part of a global network of flood-prone cities and regions, all of which are connected by Dutch hydraulics and engineering. Rotterdam is bearing the brunt of rising sea levels and heavy rainfall—we learn that the Germans "have long since raised their river dykes to funnel the water right past them and into the Netherlands … [it] will keep moving downriver to Rotterdam and the closed surge barriers" (45). Meanwhile, the narrator has "been lent out to work on the Venice, London, and Saint Petersburg surge barriers," and the Mekong Delta and New Orleans are also engineered by the Dutch (45, 46).

In stark contrast to the outsider role of Hauke Haien as the dykemaster in *Der Schimmelreiter*, Cato tells the narrator that he is "the equivalent of an atomic physicist in technological prestige" (55). The superstar status of hydraulic engineers reflects the value placed on technical solutions that characterizes this recognizable future society. His fellow citizens are not suspicious of technological progress; they believe in it utterly, to the point that there is no willingness to confront the inevitable disaster that is ahead. Technology was developed to protect Rotterdam following the terrible 1953 flood, and it will save them now too. Even the engineers themselves are refusing to believe openly in defeat, despite the evidence to the contrary.

> Ten years ago we needed to conduct comprehensive assessments of the flood defenses every five years. Now safety margins are adjusted every six months to take new revelations into account. For the last year and a half we've been told to build into our designs for whatever we're working on features that restrict the damaging effects *after* an inevitable inundation. [...] It has always been the Dutch assumption that we would resolve the problems facing us from a position of strength. But we passed that station long ago. At this point each of us understands privately that we're operating under a banner of lost control.
>
> (57)

Knowledge and technological advances have not brought security, but risk. As Ulrich Beck argues in his seminal book *Risk Society*, knowledge and technology have brought wealth, which gives rise to risk. Risk is a product of modernity.

> In the welfare states of the West … the struggle for one's "daily bread" has lost its urgency as a cardinal problem overshadowing everything else.... This development, however, withdraws the legitimizing basis from the modernization process, the struggle against obvious scarcity, for which one was prepared to accept a few (no longer completely) unseen side effects. Parallel to that, knowledge is spreading that the sources of wealth are "polluted" by growing "hazardous side effects."
>
> (Beck 1992, 20)

This implicitly also links the local situation in Rotterdam with the idea of a global risk society. Risks are no longer small and local, but enormous, with consequences that are felt across the world. Now that Holland is prosperous and the hardship of the constant battle with water has been forgotten, the magnitude of the risk has grown. If the technology fails, the price will be high, but most people prefer not to confront that fact; the narrator and his friend and colleague Kees have been nicknamed the Pessimists by their colleagues, and their gloomy attitude is contrasted with the narratives constructed by Cato (also nicknamed "the Optimist"), who is paid to persuade the public that the water situation is still under control. The stakes are high, but people prefer Cato's spin to the risk scenarios modelled by the engineers.

Communication between these two worldviews—optimism and pessimism—is not straightforward. The narrator's bleak predictions and references to historical floods and disasters are met by Cato with the complaint that they scare Henk, his 11-year-old son. Meanwhile, Cato's job requires her to ignore or play down the catastrophe that is approaching: "Her mandate was to showcase Dutch ingenuity, so the last thing she needed was the Pessimists clamouring for more funding because nothing anyone had come up with yet was going to work" (Shepard 2011, 52). Cato reassures Henk that the flood protection measures are sufficient—"Oh honey, it's more than enough" (53)—but, at the same time, she is planning his evacuation to Berlin. The narrator, meanwhile, has a secret fund of money that he has also not revealed to Cato. Their relationship is one of secrets and obfuscations.

Meanwhile, the narrator's mother, who lives alone and has dementia, adds a third element to the lack of communication between past and present, between pessimism and optimism. The narrator tells us, "my mother always had memory problems" (50), due to her traumatic experience of the 1953 floods, in which she and her parents saw her brothers swept away to their deaths. While the memories of this experience are recounted in detail, the elderly woman has short-term memory loss and confusion, unable to keep track of everyday events. Meanwhile, Rotterdam's technical and educated elites also seem to be suffering from a kind of memory loss, and an unwillingness to see the connection between the horrific floods of 1953 and the present-day threat. As long as the continuities between disaster memory and contemporary risk awareness remain severed, there is little chance of the risk of flooding being taken seriously.

The lies and avoided questions of this central relationship are mirrored in the geopolitical events that drive the plot. The narrator is pessimistic about Rotterdam's resilience to flooding: meanwhile, Cato reassures the Germans that the Dutch have the floodwaters under control.

> "We're on for another simulcast tomorrow and it'll be three Germans with long faces and Cato the Optimist." […]
>
> "Except that the people you're dealing with now *want* to be fooled," I tell her.
>
> "It's not that they want to be fooled," she answers. "It's just that they're not convinced they need to go around glum all the time."
>
> (55)

The hope that technology has the solutions has given rise to a false optimism, which endures beyond the capacity of the water engineers to prevent flooding.

This creates an interesting echo with Storm's much earlier work, in which Hauke Haien is unable to persuade his contemporaries of the need for the new dyke; there, he is the outsider, whereas here, water engineers enjoy high status. Nonetheless, the gap between the actual potential of new engineering knowledge and its perception by the outside world remains. The parallel between individual human relationships and the broader relationship of society to its own risks and hazards is subsequently made explicit:

> Cato always maintained that when it came to their marriage, her parents practiced a sort of apocalyptic utilitarianism: on the one hand, they were sure that everything was going to hell in a handbasket, while on the other they continued to operate as if things could be turned around with a few practical measures.
>
> But there's always that moment in a country's history when it becomes obvious that the earth is less manageable than previously thought.
>
> (56)

There are brief moments of unity between the protagonists of the story. The sexual relationship between Cato and the narrator provides a plane for communication in which the two of them can reach each other honestly. Sex has an almost mystical function; it is what has brought the two of them together and produced their son, Henk. What this perhaps suggests is that this, the most embodied level of human communication, is the key to the tenacity of the human species, rather than the rational human who engineers automated surge barriers and successful public relations strategies. To put this another way, humans have by and large become detached from nature—in their role as thinkers and engineers, they have lost the symbiotic relationship with their environment that in the past kept a kind of balance between the human population and the high-water mark. Sex is one of the few vestiges of animal nature that humans still have; in the technology-controlled urban setting of Rotterdam, do humans perhaps need to revert to a more "natural" or "animal" way of inhabiting the world in order to be able to understand and minimize its risks? Certainly, in both Theodor Storm's *Schimmelreiter* and the dyke-expert narrator in its modern-day counterpart "The Netherlands Lives with Water," technological advances do not seem to hold the answers.

Dyke-lines

On a thematic level, both works are ambivalent about the role of technical progress. But they also question the idea of linearity on a structural level, and this subtly undermines further the idea of humanity's striving for technical solutions to old problems. "The Netherlands Lives with Water" seems to be playing with the idea of the convergence of many different timelines. In Shepard's text, several different historical moments are invoked, all of which seem to be converging on the disaster that is about to occur in Rotterdam.

The importance of disaster memory, as explored in Chapter 1, is present in both of these texts too. In Shepard's text it is made particularly explicit; the narrator obsessively lists medieval chronicles and early modern accounts of flooding disasters:

> Bishop Prudentius of Troyes wrote in his annals that in the ninth century the whole of the country was devoured by the sea; all the settlements disappeared, and the water was higher than the dunes. In the Saint Felix Flood, North Beveland was completely swept away. In the All Saints' Flood, the entire coast was inundated between Flanders and Germany. In 1717 a dike collapse killed fourteen thousand on Christmas Night.
>
> (42)

A more recent flood memory entwined with the lives of the characters in the story, and is part of the living memory of the narrator's mother: the 1953 flood, when a tidal surge on the night of 31 January redrew the map of the Netherlands, sweeping away and drowning close to 2,000 people. This disaster was the catalyst for better warning systems around the North Sea basin, including the Thames Barrier in England and the Delta Works in the Netherlands, but it also supplies the protagonists with a direct connection to formative cultural and national memories. The pessimistic narrator, reciting his way through historic disasters, is the link between the timescales and the current moment, in contrast to most of his peers.

There is another timescale too which makes an appearance in the accounts of the area's history, and that is prehistory. We learn that crushed shell deposits can be found "seven meters higher on the dune lines inland than anyone believed floods had ever reached, deposits that look to be only about ten thousand years old" (58). The narrator also explains that "Five thousand years ago it was possible to walk from where I live to England. At that point the Thames was a tributary of the Rhine" (60). These dates in human prehistory reflect geological changes to the landscape that happened thousands of years before human recorded history, and are thus traditionally the domain of geologists, rather than historians or storytellers.

The geological processes that have deposited shells in 's-Gravensande, the medieval chronicles of storm floods, the individual memories of the flood tragedy of 1953, and the future-time of the novel for the current reader are interspersed in the short story and they overlap with and interrupt one another. The landscape of the Netherlands is shown to be a dynamic site of change, of regular floods and human responses. The effect is anti-linear; we are confronted with cyclical processes of disaster and reaction, of land reclamation and inundation, which are at odds with the linear nature of technological progress.

Der Schimmelreiter, by contrast, is structured as a chronological narrative of Hauke Haien's life from a small boy to his death on the dykes in a storm, but, despite this, it also evades simple linearity. The story starts, not with Hauke Haien, but with the first-person narrator, who recalls a story he read sat at his grandmother's knee as a child. This is the first narrator, whose words open the novel: the story he remembers reading recounts in the first person the memories of a man travelling along the dyke in high winds, who sees the mysterious apparition of the "Schimmelreiter"—the rider on the white horse. This traveller, who encounters the ghost of Hauke Haien (and thus also introduces the idea of disaster memory, since the apparition on the dyke is known for appearing during storms of particular magnitude), is the second narrator. In a nearby tavern, he tells us, he hears the story of the *Schimmelreiter* from the old schoolmaster. The schoolmaster, who is the third narrator, often completely recedes as a character, as the story of Hauke Haien unfolds as a simple third-person account: but, at key moments, he interrupts to sip his drink or move location, reminding the reader of the frame and breaking with the impression of an omniscient, watertight, disembodied narrator. In narratological terms, the schoolmaster is a hybrid, both intradiegetic to the story told by the second narrator (and thus also by the first), and extradiegetic to the story of Hauke Haien. The second narrator has the same function—he is intradiegetic in terms of the book remembered by the first narrator, but himself only functions to record the story of Hauke Haien as it is told to him. These nesting frames have a temporal function too. The first narrator is speaking to us from a position of immediacy, from the time, we assume, of the novel's publication in the late nineteenth century; but his memories of reading are set back "vor reichlich einem halben Jahrhundert" ("a good half century") (Storm 1888, 1; 1964, 184) from his account itself. The second narrator opens with:

> Es war im dritten Jahrzehnt unseres Jahrhunderts, an einem Oktober-Nachmittag—so begann der damalige Erzähler—als ich bei starkem Unwetter auf einem nordfriesischen Deich entlang ritt. Zur Linken hatte ich jetzt schon seit über einer Stunde die öde, bereits von allem Vieh geleerte Marsch, zur Rechten, und zwar in unbehaglicher Nähe, das Wattenmeer der Nordsee…
>
> (It was in the third decade of our century, on an October afternoon (so the account in the magazine began), and I was riding along a dike in Northern Friesland during extremely foul weather. For more than an hour I had seen, on my left, the desolate marsh, already empty of livestock, and on my right, uncomfortably near, the shoals of the North Sea.)
>
> (2; 184)

This tells us clearly that the second narrator is also remembering a different time, whilst himself writing from a period in time (the 1820s) almost 70 years earlier than the first narrator. While the first narrator is connected to the past through his grandmother, over 80, the second narrator is told the story of Hauke Haien by an ancient schoolmaster, who opens his own tale with the words, "In der Mitte des vorigen Jahrhunderts, oder vielmehr, um genauer zu bestimmen, vor und nach derselben, gab es hier einen Deichgrafen …" ("In the middle of the last century, or a little earlier or later, to be exact, there lived in this place a dykemaster…") (10; 188). The schoolmaster claims to be telling a rational tale, leaving out the elements that others would exploit in a ghost story, but supernatural interpretations of the story persist in his narrative despite this. As Kate Rigby notes, "Hauke is *both* the malign ghost … *and* the enlightened dyke builder" (2015, 99). The thematic tension between the supernatural and rational interpretations of the story are upheld by the subjective and fallible narrators, which Rigby refers to as "the unreliable narrator, amplified to the power of three" (98). Each of these narrators represents a perspective that allows us to look further back in time; the framing narratives present us with a wobbly, porous line of succession from Hauke Haien to the first narrator (who sounds like he might be Theodor Storm himself).

Each of these narratives is folded into the ones before, like the rings of encircling dykes on the Frisian coast. But at key moments in the text, the second narrator breaks back in, reminding us of the constructed nature of the narrative and its multiple framing devices, and completing the analogy with the dykes of the story, which are also not impermeable. The second narrator—the traveller in the storm—returns at the end of the text, but the very first, authorial narrator, remembering where he first read this story, does not. The circle is not complete; there is the final, unmendable breach in the dyke. Neither the linear trajectory of human progress, nor the encircling protection of the dyke, is complete; both are deeply entangled with the human nature that called them into being.

The convergence and entanglement of different linear models in Jim Shepard's short story and the incomplete frames of Storm's novella both underline the tension between the rationality that calls for technical progress and a linear narrative on the one hand, and the blind spots, memories, and imaginings that make up human knowledge on the other.

Nature and agency

These two works are set on parts of the coast that have a historic relationship with the hazards of sudden floods and with human attempts to manage the landscape. The representations of this landscape differ greatly between the two works. In Storm's *Schimmelreiter*, the human inhabitants of the coastline are relatively few and are clustered in largely rural settlements. The wide, flat spaces of land and sea and mudflat put the natural environment on show. Storm employs a broad, distanced perspective, as in landscape paintings (a feature that echoes his use of narrative frames); the human inhabitants serve to give a sense of scale, and perhaps awe:

> Auf der weiten Weidefläche, die sich zu Osten an der Landseite des Deiches entlangzog, sah man am Nachmittag darauf eine dunkle Menschenmasse bald unbeweglich still stehen, bald, nachdem zweimal eine hölzerne Kugel aus derselben über den durch die Tagessonne jetzt von Reif befreiten Boden

hingeflogen war, abwärts von den hinter ihr liegenden langen und niedrigen Häusern allmälig weiterrücken.... Aus den gefrorenen Gräben, welche allmälig überschritten wurden, funkelte durch die scharfen Schilfspitzen der bleiche Schein der Nachmittagssonne, es fror mächtig.

(The next afternoon, a dark mass of people milled around the wide pasture that ran along the land side of the dike towards the east. The onlookers stood still; then, after a wooden ball flew twice out of the crowd and skimmed over the earth that now lay free of frost in the sunlight, they moved slowly forward from the long, low houses that lay behind them.... The whole crowd moved across one frozen ditch after another, and the pale noon sun glittered among the sharp points of the reeds. The weather was rapidly freezing.)

(59; 211)

The Frisian lowlands are a central element of the novella, both aesthetically, in a written representation of wide landscape canvases and storms that incorporate elements of the sublime ("dunkelbraune Wolken jagten überhin, und Schatten und trübes Licht flogen auf der Erde durcheinander"; "dark brown clouds blew past, and shadows alternated with dim light over the fields" (201; 274)—compare this with, for example, Johan Christian Dahl's 1832 painting of a shipwreck off the Norwegian coast, or J.M. Turner's various stormy seas), and also in terms of the plot. The part of North Frisia where *Der Schimmelreiter* takes place has been discussed in several recent studies of Storm's work, and is also a model region in a current literary mapping project (Barz 2000; Eversberg 2014; Winkler 2012). In her post in the online literary atlas, Kathrin Winkler argues that Storm's natural places also have agency—nature is a protagonist:

Das Nebeneinander von Land und Wasser provoziert eine Konfrontation von Mensch und Natur; der Küstenverlauf schreibt zwar vermeintlich den Grenzverlauf der Bereiche fest, doch diese Grenze ist hart umkämpft. Der Mensch ringt dem Meer Land ab, das Meer nimmt es sich zurück: Der von Hauke Haien eingedeichte Koog wird schlussendlich vom Wasser "zurückerobert."

(The coexistence of land and water provokes a confrontation between humans and nature; the line of the coast would appear to mark the border between the two areas, but this border is hotly contested. Humans wrestle land from the sea, the sea takes it back: Hauke Haien's newly diked lands will, at the end of the day, be "reclaimed" by the water.)

(2012, translation my own)

The agency (or lack thereof) of the non-human environment has long been a matter of interest to ecocritics. However, there is a difference between the way nature and its agency are seen in *Der Schimmelreiter* and more contemporary representations of nature, for example the flood menace that unseats the protagonists in Jim Shepard's "The Netherlands Lives with Water." In Storm's text, the landscape is still ultimately seen as something fairly static. We do occasionally catch a glimpse of a landscape that is changing over time, as here:

> Von der Hofstelle des Deichgrafen … sah man derzeit … eine kleine Hallig, die sie "Jeversand" auch "Jevershallig" nannten. Von den derzeitigen Großvätern war sie noch nur Schafweide benutzt worden, denn Gras war damals noch darauf gewachsen; aber auch das hatte aufgehört, weil die niedrige Hallig ein paar Mal, und just im Hochsommer, unter Seewasser gekommen und der Graswuchs dadurch verkümmert und auch zur Schafweide unnutzbar geworden war.
>
> (North of the dikemaster's farm, one could see, a few thousand feet out into the shallows and somewhat farther from the opposite bank, a small island called "Jeverssand" or "Jevershallig." It had once been covered with grass and used as a sheep pasture. But the practice had been given up, because the low island had been flooded several times by the sea, particularly in the middle of summer, and the grass damaged and rendered unfit for the sheep.)
> (112; 235)

But these moments are scanty. The changes are small disruptive incidents, the result of extreme weather conditions; there is no sense that the landscape is, and always has been, permanently in flux. The only major change in the landscape that takes place during the course of the novella is the building of the new dyke and the reclamation of the Hauke-Haien-Koog, which we know exists generations after Hauke Haien's death in the storm. The Enlightenment trajectory of human progress allows for human agency and change in a largely static, if not actually passive, natural environment.

It's true that the storm, as Winkler notes above, has an element of almost-human potency and will, exacting a kind of cruel justice in the story of Hauke Haien—but seeing nature as an actor in a human story still ultimately subordinates it as a part of human imagination. Nature is not, here, a greater force than the human imaginative effort to contain it. If nature seems to be cruel, or cunning, then it is perhaps acting out the remnants of the magical qualities it was imbued with by pre-Enlightenment humanity rather than the true agency that might render it distant or unknowable from the perspectives of its human species.

The representation of the natural world in "The Netherlands Lives with Water" provides a contrast to *Der Schimmelreiter* in many ways. Shepard's future Rotterdam is the opposite of Storm's rural and sparsely populated North Frisia. The landscape has been replaced by a modern cityscape, full of human inhabitants and the trappings of modern, technological coastal society—offices, cranes, dredgers, muddy pavements. Nature has retreated into human-made channels and reservoirs, largely out of sight; the narrator has to travel out of town to check the dykes. The threat of water has become almost invisible, apart from the constant rain. The water from the sea is held back by the Delta Works while the Rhine is contained by high dykes, with signs forbidding access to civilians:

> … pedestrians are banned from many of the sea-facing dikes in the far west even on calm days. At the entrance to the Harlingvlietdam they've erected an immense yellow caution sign that shows two tiny stick figures with their arms raised in alarm at a black wave three times their size that's curling over them.
>
> (Shepard 2011, 57)

The underwater reservoirs are full to bursting and the increasingly dire measurements of the hydrological engineers are the only clear signs of the impending cataclysm.

The environment here is also characterized, not by any stasis, but by its extreme mutability. Landscape instability is not a new phenomenon; the narrator takes pains to give examples of the way that the natural world has changed over the *longue-durée*. As we have already heard, "five hundred thousand years ago it was possible to walk from where I live to England. At that point the Thames was a tributary of the Rhine" (60). The enormity of this change is obvious to anyone familiar with the map of Europe (this is an interesting scientific echo of the folk legend recorded in Hans Christian Andersen's *De to Baronesser*, explored in Chapter 1, concerning the flood that "broke through" the land bridge between England and France—see Andersen 1870, 121). In other passages, the narrator tells us of settlements destroyed by the sea and of entire regions of the country swept away. The narrator's understanding of the landscape as one in constant dialogue with the sea is strengthened by the examples of the historic floods. As the narrator observes, the need for humans to interact with their natural world was an imperative part of Dutch nation-building: "We had cooperative water management before we had a state" (Shepard 2011, 41). The Netherlands was formed not by natural land borders—the subsiding land is too unstable for that—but by a cultural response to a hazardous and changeable landscape.

The question of nature having agency here would seem more pertinent than in Storm's 1888 novella. The nature that drives events in Rotterdam no longer seems to have any of the magical or malevolent potency that brought about the end of Hauke Haien and the old dyke. Here, nature is vast, uncaring, unknowable—a vast force that is far beyond being ascribed any human-like characteristics. If nature has agency, it is on a scale that is beyond human ken.

Enlightenment

The Anthropocene has been brought about by a human belief in progress, driven by the exploitation of technology, in particular fossil-fuel technology, to power global prosperity and trade. Neo-materialists, however, hold that humanity has never been the architect of its material world, and that the Enlightenment ideal of progress and dominion was a fallacy. In his essay "Against the Anthropocene: A Neo-Materialist Perspective," Tim LeCain argues against the term Anthropocene for its centring of humanity in what he sees as a more potent combination of material processes and flows:

> Humans have obviously become quite powerful—powerful enough to populate nearly every habitable niche on the planet and alter its fundamental biogeochemical cycles. But rather than crediting humans alone, neomaterialism suggests that they accomplished these things only at the price of throwing their lot in with a lot of other things, like coal and oil, whose powers they only vaguely understood and certainly did not really control. Likewise, once these partnerships were made, these powerful things began to shape humans and their cultures in all sorts of unexpected ways, many of them not necessarily for the better.
>
> (2015, 4)

The human agency implied by the term Anthropocene, he argues, allows for the possibility of "the good Anthropocene"—of technical fixes to the heedless exploitation and overconsumption of modern-day human societies. The good Anthropocene, rather like Hauke Haien's plans for the new dyke, suggests that no fundamental allowances need to be made for nature's agency, only that ever more sophisticated human engineering projects must be devised to protect our modern way of life. Just as Hauke Haien's fanatical belief in the new dyke came at the cost of heeding the unpredictable nature of the hazardous landscape, most of the characters in Jim Shepard's short story would rather believe in technical progress than face up to the crisis of lost control.

In his "Four Theses," Dipesh Chakrabarty asks, "has the period from 1750 until now been one of freedom or that of the Anthropocene? Is the Anthropocene a critique of the narratives of freedom?" (2009, 210). Chakrabarty's concept of freedom is closely linked to the Enlightenment idea of reason; it is the emancipation of (Western) human life of scarcity, misery, and subordination to prosperity, equality, and comfort, a narrative of progress. Tim Clark suggests that:

> environmentalist thinkers disassociate themselves from unqualified endorsement of another defining aspect of the enlightenment tradition. They must divorce the project of human liberation from that of the exploitative conquest of nature that has mainstream conceptions of "progress" and "modernisation." Can the two strands of enlightenment heritage be coherently separated?
>
> (2011, 102–3)

Those concerned with environment, with the natural world in the broadest sense, are thus charged with asking Chakrabarty's question and identifying the tensions that exist both now and historically between human progress and the degradation of the natural world. While fossil-fuel energy and urbanization are dominant strands in this historical narrative, the reclamation and securing of land from the sea are also important parts of it. The advent of reason that drives Hauke Haien to build new dykes is married with the advances in technology in other areas and results in the kind of technologized, urbanized settlement that is Rotterdam in the twenty-first century—a highly populated city beset by sinking land and rising water levels, ever more dependent on technical features to warn of and prevent its inundation.

Two slight works of literary fiction can only give a limited perspective on the perceptions of these processes of technologization and risk management in the lowlands of the North Sea coasts, but dealing as they do with protagonists who are directly implicated in the engineering of these flood-prone environments, they can shed light on the tension ascribed by Chakrabarty to the struggle between progress (freedom) on the one hand, and the costs of the Anthropocene on the other.

"Shedding light on…" something is of course linked semantically and metaphorically to the idea of the Enlightenment: the idea of light as truth and darkness as ignorance. In Storm's *Schimmelreiter*, light seems to have a hard time breaking through the fog and darkness; the lack of light at the beginning of the novella makes it hard for the narrator to discern his surroundings:

> zwar sollte man vom Deiche aus auch Halligen und Inseln sehen können; aber ich sah nichts als die gelbgrauen Wellen … dahinter wüste Dämmerung, die Himmel und Erde nicht unterschieden ließ; denn auch der halbe Monde, der jetzt in der Höhe stand, war meist von treibenden Wolkendunkel überzogen.
>
> (I should have been able to look out from the dike and see small sandbars and islands; but in fact I saw nothing but yellowish-gray waves.… Behind the waste of twilight I could not tell the earth from the sky, for even the half-moon, now at its height, was almost covered by driving, cloudy darkness.)
>
> (Storm 1888, 2; 1964, 184)

The obscured view of the landscape that sets the scene for the *Schimmelreiter* is a recurring theme. Hauke Haien has a clear view of the land in the "goldene Septembersonne" ("golden sunlight of September") when he first has the idea of constructing the new dyke and reclaiming 1,000 acres (*demat*) of land (104; 231), but on the day of the great storm tide, "Schatten und trübes Licht" ("shadows alternated with dim light") (201; 274) chase each other across the sky. Hauke Haien sees clearly only in brief moments when the moon breaks through the clouds of the "furchtbaren Dämmerung" ("the terrible twilight") (208; 277). Roebling sees nature as a force working against the human intellect: "Die mit der wilden Natur verbundenen Dunkelheitsbilder erscheinen beinahe wie die Verbildlichung einer elementaren Gegenkraft zur aufklärungsgeleiteten Wahrnehmung" ("The images of darkness, aligned with nature's savagery, seem almost an elementary opposing force to Enlightenment-guided [human] perception") (2012, 321). As Hauke Haien rides to his death towards the breached dyke, the solitary beam of light comes from his home, safe on the high *Warft*; the light of truth shining bleakly out of a house devoid of people into the surrounding darkness.

In "The Netherlands Lives with Water," the question of light has become more complicated. Sources of light are artificial, as here, where light is the by-product of the Caltex refinery in the "petrochemical sprawl" of the narrator's home suburb of Pernis:

> My sister loved the lights of the complex at night and the fires that went hundreds of feet into the air like solar flares when the waste gases burned off.... The light was that golden sodium vapor light, and my father liked to say it was always Christmas in Pernis. At night I was able to read with my bedroom lamp off.
>
> (Shepard 2011, 41)

When the lights go out suddenly during the final storm, the narrator sits alone, his only source of light "the halogen emergency lights in the corridors ... with their irritated buzzing" (64). Light is no longer an unambiguously positive force that allows clearer sight and understanding, but a manufactured product tainted by the technology that has produced it. The scholar Markus Wilczek (2014) has suggested, in reference to aesthetic representations of radioactivity as light, that the truth itself is now contaminated; that the Anthropocene has poisoned the light of truth.

I now return to the question posed at the beginning of this section: "has the period from 1750 until now been one of freedom or that of the Anthropocene? Is the Anthropocene a critique of the narratives of freedom?" Storm, writing in 1880, but looking back in time to the eighteenth century, gives an ambiguous portrait of the early years of the ambitious new projects to reclaim land on the Frisian coasts. By aligning his fallible protagonist, Hauke Haien, with the humanist values of empiricism, science, and progress, Storm problematizes these values, and shows the difficulty of uniting them with a worldview rooted in spiritual and superstitious beliefs. The new dyke survives at the expense of the old one; new technology comes at the expense of old certainties. There is, though, in Storm's text, no sense of a longer-term problem with the engineering of the natural world for human needs. Only the narrative structure of nesting stories, in which the opening narrator never returns to "close" the novella fully, suggests an anti-linear subversion of the optimistic trajectory of human progress, and thus lays the foundation for an understanding of the early Anthropocene as a divergence of human and natural symbiosis.

Jim Shepard's short story "The Netherlands Lives with Water" was written at a moment in time when the critical extent of human exploitation of the natural parameters for existence was well known. The city of Rotterdam is a cipher for the Anthropocene; there is no natural feature that has not been engineered for the security and convenience of the human inhabitants. The project of freedom—progress—has started to clash with the natural world; science and technology is no longer able to solve the problem of inundation in the lowlands, and human existence there is threatened. The light of truth that seemed to be guiding the way towards a more enlightened humanity appears now only in the form of chemical artifice.

These texts, which show the unfolding and the blooming of the Anthropocene on the edge of the North Sea, are critiques of the narratives of freedom. On a thematic, structural, and symbolic level, they show the tension between Enlightenment ideas of progress for humanity and the costs that such progress can incur. The low-lying terrain of the North Sea coast is shown to be a frontier between human and non-human forces; one that stretches across time, as well as through space.

Notes

1 English versions date back to the early twentieth century: a translation entitled "The Rider on the White Horse" was published in volume 11 of a series called The German Classics in 1914; the most recent translation is from 1996 by Denis Jackson, entitled *The Dykemaster*.
2 There is a misapprehension that medieval dykes were steep and that the sloping dykes were introduced later, after catastrophic storm floods; the development of dykes was in fact more complicated (Mauelshagen 2009, 51–2; Eversberg 2014, 429–32).
3 All German quotations from *Der Schimmelreiter* in this chapter are from Storm (1888). All English quotations are from the translation by James Wright (Storm 1964).
4 "Couldn't" (Storm 1964, 232) here is a mistranslation, and "shouldn't" would be a better rendering of the German: "… von Kindesbeinen an hab' ich gehört, der Priel sei nicht zu stopfen, und darum dürfe nicht daran gerührt werden."
5 See also Rieken on the religious interpretation of extreme events: "Weil die religiöse Deutung des Flutgeschehens als Strafe Gottes sowie magische Vorstellungen Ausdruck ein und derselben Geisteshaltung sind … ist es völlig abwegig zu behaupten, dass sich im Gefolge der Aufklärung rationale Erklärungen der Sturmfluten auf breiter Basis bereits bis zum Ende des 18. Jahrhunderts durchgesetzt hätten" ("Since both the religious attribution of flood events to divine punishment and the imagination of them as supernatural are both expressions of the same mode of thought … it is completely false to assert that the Enlightenment had already led to a widespread adoption of rational explanations for storm floods before the end of the eighteenth century"; translation my own.) (2005, 286).
6 The large-scale government project "Ruimte voor de Rivier" (Room for the River) links up a range of measures along Dutch rivers to increase safety and reduce the risk of high water by, for example, moving dykes outward and widening floodplains. www.ruimtevoorderivier.nl/english/.

References

Andersen, Hans Christian. 1870. *The Two Baronesses: A Romance*. Translated by Charles Beckwith Lohmeyer. New York: Hurd & Houghton.

Arndt, Christiane. 2012. "Theodor Storm's *Der Schimmelreiter: Schauerrealismus* or Gothic Realism in the Family Periodical." In *The German Bestseller in the Late Nineteenth Century*, edited by Charlotte Woodford and Benedict Schofield, 144–62. New York: Camden House.

Bankoff, Greg. 2013. "The 'English Lowlands' and the North Sea Basin System: A History of Shared Risk." *Environment and History* 1: 3–37.

Barz, Paul. 2000. *Der wahre Schimmelreiter: Die Geschichte einer Landschaft und ihres Dichters Theodor Storm*. Hamburg: Convent.

Beck, Ulrich. 1992. *Risk Society: Towards a New Modernity*. Translated by Mark Ritter. London: Sage.

Blackbourn, David. 2007. *The Conquest of Nature: Water, Landscape and the Making of Modern Germany*. London: Pimlico.

Chakrabarty, Dipesh. 2009 "The Climate of History: Four Theses." *Critical Enquiry* 35 (2): 197–222.

Clark, Timothy. 2011. *The Cambridge Introduction to Literature and the Environment*. Cambridge: Cambridge University Press.

Colten, Craig. 2012. "Forgetting the Unforgettable: Losing Resilience in New Orleans." In *American Environments: Climate–Culture–Catastrophe*, edited by Christof Mauch and Sylvia Mayer, 159–76. Heidelberg: Universitätsverlag Winter.

Demandt, Christian. 2010. *Religion und Religionskritik bei Theodor Storm*. Berlin: Erich Schmidt Verlag.

Eversberg, Gerd, ed. 2014. *Der Schimmelreiter: Novelle von Theodor Storm. Historisch-kritische Edition*, Berlin: Erich Schmidt Verlag.

Frühwald, Wolfgang. 1981. "Hauke Haien, der Rechner: Mythos und Technikglaube in Theodor Storms Novelle 'Der Schimmelreiter.'" In *Literaturwissenschaft und Geistesgeschichte: Festschrift für Richard Brinkmann*, edited by Jürgen Brummack, 438–57. Tübingen: Max Niemeyer.

Kazmierczak, Aleksandra and Jeremy Carter. 2010. "The Netherlands Lives with Water: Public Awareness Raising Campaign." In *Adaptation to Climate Change Using Green and Blue Infrastructure: A Database of Case Studies*, 19–25. Manchester: University of Manchester. eScholarID:128518.

Kimmelman, Michael. 2013. "Going With the Flow." *New York Times*, 13 February 2013. www.nytimes.com/2013/02/17/arts/design/flood-control-in-the-netherlands-now-allows-sea-water-in.html (last accessed 24 March 2016).

LeCain, Timothy J. 2015. "Against the Anthropocene: A Neo-Materialist Perspective." *International Journal for History, Culture and Modernity* 3 (1): 1–28.

Mauelshagen, Franz. 2007. "Flood Disasters and Political Culture at the German North Sea Coast: A Long-term Historical Perspective." *Historical Social Research* 32 (3): 133–44.

Mauelshagen, Franz. 2009. "Disaster and Political Culture in Germany Since 1500." In *Natural Disasters, Cultural Responses: A World History*, edited by Christof Mauch and Christian Pfister, 41–75. Lanham, MD: Lexington Books.

Mullen, Inga E. 1988. *German Realism in the United States: The American Reception of Meyer, Storm, Raabe, Keller and Fontane*. New York: Peter Lang.

Rieken, Bernd. 2005. *Nordsee ist Mordsee: Sturmfluten und ihre Bedeutung für die Mentalitätsgeschichte der Friesen*. Münster: Waxmann.

Rigby, Kate. 2015. *Dancing with Disaster: Environmental Histories, Narratives, and Ethics for Perilous Times*. Charlottesville: University of Virginia Press.

Ritson, Katie. 2017. "Engineering the Anthropocene: Technology, Ambition, and Enlightenment in Theodor Storm's *Der Schimmelreiter*." In *Readings in the Anthropocene: The Environmental Humanities, German Studies, and Beyond*, edited by Sabine Wilke and Japhet Johnstone, 222–42. New York: Bloomsbury.

Roebling, Irmgard. 2012. *Theodor Storms ästhetische Heimat: Studien zur Lyrik und Erzählwerk Storms*. Würzburg: Königshausen & Neumann.

"Ruimte voor de Rivier" ("Room for the River"). www.ruimtevoorderivier.nl/english/ (last accessed 16 August 2018).

Shepard, Jim. 2011. "The Netherlands Lives with Water." In *You Think That's Bad: Stories*, 38–66. New York: A.A. Knopf, 2011.

Storm, Theodor. 1888. *Der Schimmelreiter*. Berlin: Verlag der Gebrüder Paetel.

Storm, Theodor. 1914. "The Rider on the White Horse." In *The German Classics* vol. 11, 225–345. New York: The German Publishing Company.

Storm, Theodor. 1964. *The Rider on the White Horse and Selected Stories by Theodor Storm*. Translated by James Wright. New York: Signet.

Storm, Theodor. 1996. *The Dykemaster*. Translated by Denis Jackson. London: Angel Books.

Sullivan, Heather I. 2012. "Dirt Theory and Material Criticism." *Interdisciplinary Studies in Literature and Environment* 19 (3): 515–31.

van Dam, Petra. 2016. "An Amphibious Culture: Coping with Floods in the Netherlands." In *Local Places, Global Processes*, edited by Peter Coates, David Moon, and Paul Warde, 78–93. Oxford: Oxbow Books.

van Tielhof, Milja. 2015. "Forced Solidarity: Maintenance of Coastal Defences along the North Sea Coast in the Early Modern Period." *Environment and History* 21 (3): 319–50.

Wilczek, Markus. 2014. "When Our Future Is at Stake: A Poetics of Radioactivity." Talk given at the Rachel Carson Center Lunchtime Colloquium on 31 June 2014. www.youtube.com/watch?v=GElzwXx2xsE (last accessed 24 March 2016).

Winkler, Kathrin. 2012. "Nordfriesland—Der Raum als Akteur." In *The Literary Atlas of Europe: Mapping and Analysing the Geography of Fiction with Interactive Tools*. www.literaturatlas.eu/2012/02/16/northern-frisia-space-as-actor/ (last accessed 24 March 2016).

3 A sense of place in the Anthropocene

W.G. Sebald and East Anglia

The North Sea lowlands as a temporal and spatial nexus is the concern of W.G. Sebald's *Die Ringe des Saturn: Eine englische Wallfahrt* (the title translates as *The Rings of Saturn: An English Pilgrimage*). Sebald's novel-length account of a journey on foot through Suffolk examines the situation of the English lowlands in East Anglia as both a physical locality and a player in a network of ecological and cultural interactions that span the globe and all of human history. Many scholars, including Dipesh Chakrabarty, have pointed to the problematic inequalities of the Anthropocene and the ways that actions in one place and time can have unforeseen consequences in distant and future places. Sebald's portrayal of these injustices and his linking of the English North Sea lowlands to global processes of exploitation, fossil fuel use, species loss, pollution, human conquest, and geological erosion, allow this text to be read as a literary account of the Anthropocene that has the marginal English territory at its heart.

Die Ringe des Saturn was first published in Germany in 1995 and followed by an English translation (with the title shortened to *The Rings of Saturn*) by Michael Hulse in 1998. W.G. Sebald has defied allegiance to any single literary genre and to a single literary culture: he is frequently referred to as a German-British writer (e.g., Goodbody 2014a, ii) and an "England-deutscher" (Catling 2008), although he wrote only in German. His works are at least as well known in the English-speaking world through their carefully wrought translations by Hulse, Michael Hamberger, and Anthea Bell as they are in their original German, and Sebald's works show increasing influence in literary scholarship across the world. *The Rings of Saturn* can be described as a novel, "prose narrative" (Weston 2011, 172), travel writing (Macfarlane and Solnit 2014),[1] memoir, and autobiographical fiction: it has recently begun to be labelled also as an Anthropocene text (Malkmus 2017). To confuse the issue of genre further, Sebald intersperses his texts with images—grainy, black-and-white images that are not labelled or attributed—which encourage an engagement with the visual, with the nature of reproduction, and the relationship between images and words.

The Rings of Saturn (for convenience here, I will refer to the work by its shorter English title, while citing from the original German text together with

Michael Hulse's translation) has as its basic plot a first-person account of a walking tour through Suffolk. The narrator—who sounds very much like the author himself—uses this slow and meandering trip as an opportunity for an equally meandering and circuitous tour of moments in human history, using what Colin Riordan has described as "the Sebaldian techniques of history by association, of minute exegesis of the by-ways of the past, of moving through time by moving through space" (Riordan 2014, 45). Although the historical reach of the novel is global and chronologically wide-ranging—from sixteenth-century Western thought to China during the Taiping rebellion, from the Holocaust to Congo under Belgian rule—the East Anglian walk as a plot device is far from being simply arbitrary, or a convenience choice based on Sebald's biography (he lived in neighbouring Norfolk from 1970 until his death in 2001). The Suffolk coast is an integral part of Sebald's exploration of historical patterns and processes in the wider world.

Place and space in the English North Sea lowlands

Sebald's use of space and place has already given rise to a wealth of recent scholarship, including two separate (and simultaneous) literary cartographical projects (Hui 2008–09; Covindassamy 2014),[2] a film (Gee 2012), and a large number of scholarly articles, monographs, newspaper articles, and also works of homage and commemoration that link him to the places of his books (Cook 2014; McCrum 2013; Wood 2013). In this chapter, I will examine Sebald's use of space and place specifically in the broader setting of the North Sea lowlands, and the relevance this has for a discussion of human history during the Anthropocene.

Geographers have long been interested in the difference between space and place. This question has been at the heart of the spatial turn in literary and cultural studies, which, in turn again, has brought about interdisciplinary fields of study such as literary geography (Alexander 2015, 3–6). In his 2011article "Space and Place," the geographer John A. Agnew points to the longstanding conflict in geography about the "two dominant meanings," place and space, and the tendency to exalt abstractions that he ascribes to modernism, compared with place as an "empirically useful concept" (Agnew 2011, 317, 318). Agnew presents space and place as a dialectic of modernity versus anachronism, arguing that "place is often associated with the world of the past and location/space with the world of the present and future" (319). Agnew notes that contemporary fears about a loss of place have not come to pass:

> the main current challenge to both of the dominant meanings comes from the idea that the world itself is increasingly "placeless" as space-spanning connections and flows of information, things and people undermine the rootedness of a wide range of processes anywhere in particular.… Yet, previous rounds in the diffusion of technological innovation, even though often touted as likely to do much the same thing (roads, railways,

> telegraphy, ship canals, etc.), had no such effect. What they did do was help reconstitute and reorganize spatial relations such that places were remade and reconfigured.
>
> (319)

Sebald's *Rings of Saturn* is concerned with loss, and the tension between place and space is important in this regard.

Agnew makes the point that, in recent geographical scholarship, space, representing the global and the modern, is often privileged over place. Louise Mønster (2009), in her article "At finde sted" (finding place), points to a heightened need to make room for places in the face of a globalized world in which specificities of place are being lost and human presence can even be uncoupled from physical location via technologies such as Skype. In the ecologically concerned humanities, on the other hand, place has been privileged over space in the guise of bioregionalism, and it is only relatively recently that concepts such as Ursula Heise's eco-cosmopolitanism (Heise 2008) have been employed to argue against "the cult of localism" (Clark 2011, 137). For Agnew, the explicit value in geographical explorations of place lies in overcoming the gaps between different conceptions of place, and understanding the networks and connectivities that define places (Agnew 2011, 326). Sebald's choice to base his book around a walking tour of Suffolk could thus be seen as a reaction against an increasing (perceived or real) placelessness in the contemporary world, or an expression of a bioregionalist environmentalist agenda; its global reach, however, also responds to Agnew's call to recognize connectivities between space and place, and to Chakrabarty's problematization of the multiple scales of the Anthropocene.

Sebald chose an area of the countryside that is popularly perceived in England as being somewhat backward, or at least marginal and rural. The Suffolk coastline forms the edge of the English mainland, with Sebald's route touching on the easternmost sections of the coast, where the land is rapidly succumbing to erosion by the strong tides of the North Sea Basin. Towns such as Lowestoft and the almost extinct Dunwich are presented as sites of decline, of failure to keep up with the modern world:

> In den langen Straßenzügen, durch die ich gehen mußte, zeigte sich keine lebende Seele, und je mehr ich dem Zentrum mich näherte, desto mehr bedrückte mich, was ich sah.... Unbegreiflich schien es mir darum jetzt, als ich nach Lowestoft hineinging, wie es in einer verhältnismäßig so kurzen Zeit soweit hätte herunterkommen können.... Heute steht in manchen Straßen fast jedes zweite Haus zum Verkauf, Unternehmer, Geschäftsleute und Privatpersonen versinken immer weiter in ihren Schulden, Woche für Woche hängt irgendein Arbeitsloser oder Bankrotteur sich auf, der Analphabetismus hat bereits ein Viertel der Bevölkerung erfaßt, und ein Ende der stetig fortschreitenden Verelendung ist nirgends abzusehen.
>
> (Sebald 1997, 55–6)

(Not a living soul was about in the long streets I went through, and the closer I came to the town centre the more what I saw disheartened me.... so now, as I walked into Lowestoft, it seemed incomprehensible to me that in a relatively short period of time the place could have become so run down.... Nowadays, in some of the streets almost every other house is up for sale; factory owners, shopkeepers and private individuals are sliding ever deeper into debt; week in, week out, some bankrupt or unemployed person hangs himself; nearly a quarter of the population is practically illiterate; and there is no sign of an end to the encroaching misery.)

(Sebald 1998, 41–2)[3]

Moreover, Sebald's narrator moves through the landscape on foot, adding to the anachronistic sense of seeking out a place that is more past than present. Frequently, as above, Sebald's text portrays a town or section of countryside that is empty of people—a reflection not just of the narrator's solitary travels through his own mind, but also suggestive of a local exodus of people towards modernity. Sebald even talks about "westing," the inexorable drive westwards that has been observed in human and non-human populations alike (191; 158).

But to read Sebald's exploration of Suffolk as a literary articulation of an old ideal of "place" or an environmentalist attempt at reinhabitation would be too simplistic. Sebald simultaneously uses the narrator's walk through the low-lying landscape as a means of engaging with the globalized notion of "space."

So unbegreiflich mir diese Dinge von jeher gewesen sind, so unmöglich war es mir auch an jenem Abend auf dem Gunhill von Southwold, wirklich zu glauben, daß ich vor genau einem Jahr vom holländischen Strand aus nach England hinübergeschaut hatte. Ich war damals … nach Den Haag gefahren und hatte mich dort in einem der zweifelhaften Hotels am Stationsweg einquartiert.

(Just as these things have always been beyond my understanding, so too I found it impossible to believe, as I sat on Gunhill in Southwold that evening, that just one year earlier I had been looking across to England from a beach in Holland. On that occasion … I had travelled … to the Hague, where I had taken a room in one of the less salubrious hotels near the station.)

(99; 80)

[…]

Stattdessen kaufte ich mir bei MacDonald's, wo ich mir an dem grellerleuchteten Tresen wie ein seit langem in allen Ländern gesuchten Verbrecher vorgekommen bin, eine Tüte Chips, die ich auf dem Rückweg in mein Hotel nach und nach verzehrte. Vor dem Eingängen der diversen Unterhaltungs- und Eßlokale am Stationsweg hatten sich inzwischen kleine

> Gruppen morgenländischer Männer versammelt, von denen die meisten stillschweigend rauchten.... Als ich den kleinen Kanal erreichte, der den Stationsweg kreuzt, glitt auf einmal quer über die Fahrbahn an mir vorbei, als sei sie aufgetaucht aus dem Nichts, eine mit Lichtern bestückte, chromglänzende amerikanische Limousine mit offenem Verdeck...
>
> (No longer able to decide on a place to eat, I bought a carton of chips at McDonalds, where I felt like a criminal wanted worldwide as I stood at the brightly lit counter, and ate them as I walked back to my hotel. Outside the entrances of the entertainment and dining establishments on the road to the station, small groups of oriental men had now gathered, most of them smoking in silence.... When I reached the little canal that crosses the road, an open-top American limousine studded with lights and gleaming with chrome glided past me across the carriageway as if it had come out of nowhere...)
>
> (101; 81)

In characteristic thought process, the narrator's view of the sea in Southwold reminds him of The Hague, just across the water, transgressing the national setting of the text: The Hague, in his account, shows itself to be very much a modern space (rather unlike Southwold); it is full of the signs of a globalized world—a Persian carpet shop that conceals the entrance to a mosque, a black man pursued by an ostensible compatriot wielding a kitchen knife, an American limousine. Moreover, the narrator buys himself food from a ubiquitous fast-food outlet—somewhere emblematic of the "non-place," in the language of Marc Augé. Augé writes that "If a place can be defined as relational, historical and concerned with identity, then a space which cannot be defined as relational, or historical, or concerned with identity will be a non-place. The hypothesis advanced here is that supermodernity produces non-places" (1995, 77). Sebald moves from place to non-place along a chain of transnational associations and places, not simply contrasting place with space, but linking them; his thought processes make connections and find patterns between different places in both space and time. Mandana Covindassamy argues in her book that Sebald's writing is *écriture en déplacement*—writing that constantly displaces itself (Covindassamy 2014) and Sebald's narrator is both in place and permanently moving through space.

When the narrator glimpses the entrance to the mosque, he catches sight of a pile of shoes—"mein Blick einen mir unvergeßlichen, ganz aus der Zeit gelösten Moment lang auf eine hölzerne Stellage gefallen ist, in der, ordentlich über- und nebeneinander, vielleicht hundert Paar abgetragene Straßenschuhe standen" ("for an unforgettable moment that seemed to exist outside time, I glimpsed the wooden rack on which perhaps a hundred pairs of well-worn shoes had been placed beside and above one another") (81; 100). This image, in the context of Sebald's preoccupation with human darkness, is reminiscent of the well-known photos from Auschwitz of the heaps of shoes left behind by victims of the gas chambers; a further element of Sebald's web of connections, it draws an additional historical moment into his memory of The Hague from a bench in Suffolk. Covindassamy's use of the word displacement to describe his technique aligns Sebald's narrative style with one of his overriding themes, displacement, seen here in a reference to the Holocaust and the widespread movement of people during and after World War II. Sebald's relationship with place is not simple—it is not just connective across space, but also fundamentally fragmented, dislocated. Sebald's narrator in *The Rings of Saturn* does not "inhabit" his landscape in the manner favoured by many environmentalist writers "in order to overcome our alienation from the natural environment" (Goodbody 2014b, 69); instead, he is a traveller through it.

In her 2008 book *Sense of Place and Sense of Planet: The Environmental Imagination of the Global*, Ursula Heise situates her argument about place and space explicitly in the context of the environmental movement, which burgeoned in the United States and western Europe in the 1960s and 1970s, and has become a pluralized feature of the contemporary world since then. She is critical of the texts that see the (re)inhabitation of place as the only valid form of ecological writing, insisting instead that:

> Rather than focusing on the recuperation of a sense of place, environmentalism needs to foster an understanding of how a wide variety of both natural and cultural places and processes are connected and shape each other around the world, and how human impact affects and changes this connectedness.
>
> (Heise 2008, 21)

She is critical of North American environmentalist writing in particular, asserting that environmentalist thought "has not connected to the foundational idea in much recent cultural theory that identities are at their core made up of mixtures, fragments, and dispersed allegiances to diverse communities, cultures, and places" (Heise 2008, 42–3). This would seem to be particularly relevant in the case of Sebald—more than a traveller, he is a foreigner, an outsider, his identity consciously hybrid. Jo Catling explores Sebald's projection of himself as homeless, writing of his "bewusste Außenseiterposition" ("self-conscious role as an outsider") and his interest both in stories of exile and his own sense of not "feeling at home" in England (Catling 2008, 26). His work is not local in any idealized sense, since the environmentalist fixation with the local, according to Heise, is predicated on the idea of belonging.

If Sebald's *The Rings of Saturn* can be seen as an environmentally concerned work, then, it is not (or at least not only) through its connection with a specific place, but through its imagination of an interconnected globe, in which human actors have an impact on the environment and in turn are affected by it. And, although Sebald is preoccupied most notably with the failure of human endeavours, with the collapse of civilizations and the causes and effects of human depravity and cruelty, his foregrounding of humanity does not preclude a more planetary/ecological worldview, as Colin Riordan argues in his exploration of Sebald's poetic triptych *After Nature*, which he terms "ecocentric":

> The unmissable strengths of Sebald's approach, the ability to see ourselves in context, to think in terms of connective processes, are in essence ecocentric. Sebald's narrative techniques … have as an inevitable consequence the effect that the natural world, which is the ultimate context, becomes a central object of textual enquiry.
>
> (Riordan 2014, 46)

The term ecocentric can be used in relation to *The Rings of Saturn* too: the natural world, the Earth, is the ultimate context of all human achievements and all human failures, from the birth of *Homo sapiens* to the decline and extinction of the human species. Sebald's text is human-centred, both through its narrative structure (an individual human passage through landscape and memory) and its primary focus on human events and material products (battles, enslavement, conquest—country houses, formal gardens, railway carriages), but these are firmly embedded in a wider context of the natural world and its non-human dynamics.

One such instance is Sebald's account of the fishing of herring in the North Sea basin. On passing stationary anglers on the beach south of Lowestoft, the narrator considers the dwindling fish stocks in the North Sea and the decline of professional fishermen and fishing boats on the English coast.

> Die Boote, in denen früher die Fischer von den Stränden ausliefen, sind, seit das Geschäft nicht mehr lohnt, verschwunden, die Fischer selber ausgestorben. Niemand hat ein Interesse an ihrer Hinterlassenschaft. Hie und da stößt man auf einen Schiffsfriedshof, wo die herrenlose Kähne zerfallen und die Drahtseilwinden, mit denen man sie einst ans Land gezogen hat, in der Salzluft verrosten.
>
> (The boats in which the fishermen once put out from the shore have vanished, now that fishing no longer affords a living, and the fishermen themselves are dying out. No one is interested in their legacy. Here and there one comes across abandoned boats that are falling apart, and the cables with which they were once hauled ashore are rusting in the salt air.)

The text continues by citing the causes of the declining numbers of fish:

> Tausende von Tonnen Quecksilber, Kadmium und Blei, Berge von Düngemitteln und Pestiziden werden von den Flüssen und Strömen Jahr für Jahr hinausgetragen in den deutschen Ozean. Ein Großteil der schweren Metalle und der anderen toxischen Substanzen setzen sich in den seichten Gewässern der Doggerbank ab, wo ein Drittel der Fische bereits mit seltsamen Auswüchsen und Gebresten zur Welt kommen. Immer wieder sichtet

man vor der Küste über viele Quadratmeilen sich ausdehnende, dreißig Fuß in die Tiefe hängende giftige Algenfelder, in denen die Seetiere scharenweise verenden. Einige der selteneren Schollenarten, Karauschen und Brassen, bei denen mehr und mehr der weiblichen Fische in bizarrer Mutation männliche Geschlechtsorgane ausbilden, vollführen ihre mit der Fortpflanzung verbundenen Rituale nur noch als einen Todestanz.

(Every year the rivers bear thousands of tons of mercury, cadmium and lead, and mountains of fertilizer and pesticides, out into the North Sea. A substantial proportion of the heavy metals and other toxic substances sink into the waters of the Dogger Bank, where a third of the fish are now born with strange deformities and excrescences. Time and time again, off the coasts, rafts of poisonous algae are sighted covering many square miles and reaching thirty feet into the deep, in which the creatures of the sea die in shoals. In some of the rarer varieties of plaice, crucian or bream, the females, in a bizarre mutation, are increasingly developing male sexual organs and the ritual patterns of courtship are now no more than a dance of death.)

(69–70; 52–3)

The decline of a landscape's biodiversity is given a concrete and anthropogenic cause: chemicals from human agriculture and industry have poisoned the waters; they are responsible for upsetting the North Sea ecology and for bringing about the demise of both large quantities of fish and of the fishing practices that characterized the coast. This dismal portrayal of anthropogenic environmental damage is contrasted with the history of herring fishing on the coasts, in which such an abundance of fish was available that catching, preserving, and transporting herring became an industrialized process. The little-understood process by which herring reproduced, in some years to excess, seemed to promise a nature of unlimited plenty, available for humans to harness at will. The herring was, according to Sebald's narrator, a popular object lesson in schools, promoting:

[die] Vorstellung von der staunenswerten Selbstvermehrung und Vervielfältigung des organischen Lebens, mit der wir noch aufgewachsen sind … Nicht umsonst war der Hering … das Hauptemblem für die Unausrottbarkeit der Natur.

(the notion of the wondrous increase and perpetuation of life with which we grew up. It was not without reason that the herring was … the principal emblem, as it were, of the indestructability of Nature.)

(70; 53)

The narrative of decline apparent in the shift from abundance to scarcity, from a bounteous nature to one that is almost destroyed, is however unseated when the narrator goes on to consider the unknown sentience of the fish and the possibility of their suffering at human hands, recounting cruel experiments on fish to see how long they took to expire out of water. He concludes that "in Wahrheit wissen wir nichts von den Gefühlen des Herings" ("but the truth is we do not know what the herring feels") (75; 57), leaving the reader ambivalent about whether the loss of large-scale herring fishing, which was responsible for the blooming of prosperity along the coasts, constitutes decline, or indeed on whose terms.

The possible cruelty of herring fishing is given further emphasis in two images that appear in this chapter. The first is a reproduction of an old photograph (seemingly from the early part of the twentieth century) entitled "A Morning Catch of Herring, Lowestoft" that shows a large factory hall where the floor is covered with a huge number of fish. The herring are piled perhaps a foot deep across the width of the hall, their scaled bodies lying criss-crossed over one another, while solemn and triumphant fishermen surround them in an image of conquest. Four pages on, a startling double-page spread shows a photograph of a grove of trees under which piles of human bodies have been placed. This latter image is in reference to an account of a deceased local inhabitant who was involved in the liberation of Bergen-Belsen concentration camp in April 1945; we thus assume the unattributed image shows Bergen-Belsen on the Lüneberg Heath immediately after the camp was overrun by Allied forces. The images are not in reference to one another and no link is made in the text between the two photographs, nor between the stories (other than the links to Suffolk) that they each relate to. However, the juxtaposition of the slippery pile of dead or dying herring and the piles of pale and contorted human corpses creates a disturbing visual link and raises discomfiting questions. Can we compare the suffering of herring with the suffering of humans? Are narratives of progress and narratives of destruction merely matters of perspective?

Bernhard Malkmus suggests that the deliberate weaving together of natural and cultural histories—"systematisch durchflicht Sebald Kulturgeschichten mit Naturgeschichten, Geschichten von Menschen mit Geschichten vom

Figure 3.1 Illustration from W.G. Sebald's *Die Ringe des Saturn*.

Figure 3.2 Illustration from W.G. Sebald's *Die Ringe des Saturn*.

Fisch"—is a deliberate provocation to the reader, implying the profound roots of both ecological devastation and genocide in the epistemes of modernity (Malkmus 2015–16, 194). The way that natural histories and human histories are entwined gives rise to the "natural history of destruction" that is a recurring theme in Sebald's works and the English title of his 1999 work *Luftkrieg und Literatur* (*On the Natural History of Destruction*, published in English in 2003). The landscape of East Anglia is linked both spatially, through the transnational nature of herring fishing, and historically to the Holocaust. The category slippage between the natural and the cultural is an intentional feature and allows for different perspectives on human history (as a product of human and non-human agency) to co-exist simultaneously; I will return to this point later in the chapter.

Another natural-cultural history that has a particular place in *The Rings of Saturn*, and which links the landscape that the narrator is walking through to global stories and trajectories, is that of the silkworm. The silk spun by these creatures is threaded throughout the work, linking otherwise unconnected anecdotes: Thomas Browne, we learn in the first chapter, is the son of a silk trader; in chapter six, Empress Dowager Tzu-hsi is alleged to have cared more for her silkworms than her drought-stricken subjects; a guest in Swinburne's home reports that the poet calls to mind the ashy grey silkworm, *Bombyx mori* (Sebald 1997, 198; Sebald 1998, 165). But it is not until the final chapter that the silkworm species and its place in human endeavour is explored.

The examination of the silk industry is linked to that of herring fishing; according to the narrator, he finds a German documentary film about the silkworm in his home town while searching for the documentary about North Sea fishing that he remembered from childhood (354; 291). But more than the herring, the human mastery of the silkworm represents the globalization of knowledge and trade. Sebald traces the path of silk-farming from its roots in China via Byzantine culture to Greece, Sicily, and Naples, and thence northwards to northern Italy and France. The Huguenot refugees from France who arrived in England in the seventeenth century contributed to the development of the silk industry there, in particular in Norwich, the city where Sebald lived (and where the narrator's stay in hospital provides the opening of *The Rings of Saturn*). Sebald even weaves Norwich silk into his own autobiography, commenting that "… gelangten … das eine oder andere halbseidene Hochzeitstuch vielleicht gar in der Buckelkraxe eines jüdischen Hausierers bis nach Isny, Weingarten oder Wangen" ("some half-silk wedding shawl might even reach Isny, Weingarten or Wangen in the pannier on a Jewish pedlar's back" (340; 286), towns in the west German district of Allgäu where the author was born and raised.

The threads that link Norwich across space and time via French refugees to medieval Italy and ancient China are part of Sebald's complex construction of place and space. His employment of both the extremely local and subjective places and viewpoints (Norwich, the backpack of a Jewish peddler), global connections and conceptions of space (the spread of silk farming from China to northern Europe), and both cultural and natural (ecological) processes creates a multi-perspective narrative. This is perhaps an answer to the challenge that Heise identifies for artists and writers of creating "a vision of the global that integrates allegory—still a mode that is hard to avoid in representations of the whole planet—into a more complex formal framework able to accommodate social and cultural multiplicity" (Heise 2008, 21). Sebald brings different historical and spatial scales into conversation, transcending the limitations of both "place" and "space" to create a dynamic sense of a planetary ecosystem.

The consideration of the suffering of herring and the ethical implications of killing silkworm for their thread aligns the suffering of non-human species with Sebald's exploration of human suffering at the hands of those more powerful. It also constantly undermines the idea of a narrative of progress—the beauty of silk and the abundance of caught herring, signs of human ingenuity, result from cycles of conquest and brutality. These issues are ones that have become current in discussions of environmental crisis, in which the idea of human progress and emancipation has become inextricable from the destruction of the natural world and the depletion of future resources, and the human worldview has been decentred.

So Sebald's positioning of Suffolk as a small part of a networked and globalized *space* emphasizes its connectivity across both temporal and spatial planes, linked by trade, weather, warfare, and human associations and memories with countless other places spread across the planet: and this sense of the global extends beyond human connections to ecological ones, incorporating a sense of a global nature as well as a global culture. *The Rings of Saturn* is, in Colin Riordan's term, ecocentric—it is concerned with humans within a broader ecology of non-human agency. Sebald foregrounds the human stories but embeds them in the narratives of oceans, rivers, forests, species reproduction, and coastal erosion—narratives that unfold according to their own ecological rules. The humans, like the herring and the silkworms, are nested inside the narratives of global nature, three suffering species amongst many.

The erosion of progress: East Anglia in the Anthropocene

Beyond its role as a connecting point in a globalized understanding of *space*, the uniqueness of Suffolk as a *place* underlines one of the main thematic threads of Sebald's work, the narrative of decline. The Suffolk coast is the fastest eroding coastline in Europe and its visible instability and vulnerability to storms and tides have been a constant feature of the landscape (see e.g., British Dept for Environment, Farming and Rural Affairs, 2010). Sebald writes of a tidal lagoon that is "umgeben von dem grünen Kranz eines Laubgehölzes, das aufgrund der fortschreitenden Erosion der Küste von der Meerseite her allmählich abstirbt" ("encircled by deciduous woodland that is now dying, owing to the steady erosion of the coastline by the sea") (77; 59), and dwells on the *memento mori* of the remaining stones of Dunwich:

> Das heutige Dunwich ist der letzte Überrest einer im Mittelalter zu den bedeutendsten Hafenplätzen Europas zählenden Stadt. Mehr als fünfzig Kirchen, Klöster und Spitäler hat es hier einmal gegeben, Werften und Befestigungsanlagen, eine Fischerei- und Handelsflotte mit achtzig Fahrzeugen und Dutzende von Windmühlen. All das ist untergangen und liegt, über zwei, drei Quadratmeilen verstreut, unter Schwemmsand und Schotter draußen auf dem Boden des Meeres. Die Pfarrkirchen … sind, eine um die andere, über die stets weiter zurückweichende Klippe hinuntergestürzt und nach und nach in die Tiefe versunken.
>
> (The Dunwich of the present day is what remains of a town that was one of the most important ports in Europe in the Middle Ages. There were more than fifty churches, monasteries and convents, and hospitals here; there were shipyards and fortifications and a fisheries and merchant fleet of eighty vessels; and there were dozens of windmills. All of it has gone under, quite literally, and is now below the sea, beneath alluvial sand and gravel, over an area of two or three square miles. The parish churches … one after the other, toppled down the steadily receding cliff-face and sank in the depths.)
>
> (187–8; 155)

Wherever he looks, the landscape is bare, unpopulated, and shows signs of the encroaching sea. As Axel Goodbody (2010, 134) writes:

> die Landschaften Ostenglands werden hier zu Orten des Untergangs und der Hoffnungslosigkeit, Inbildern des apokalyptischen Grauens der Endzeitlichkeit ... Dieser Verfallsprozess wird beispielsweise von der abbröckelnden englischen Südostküste versinnbildlicht, die mitsamt ihren versinkenden Küstenstädten vom Meer unterwaschen und verschlungen wird.
>
> (the landscapes of eastern England here become places of downfall and hopelessness, the essence of the apocalyptic dread of the end of days ... this process of decay is symbolized by the crumbling southeastern coast of England, which, along with its sinking towns, is being washed away and swallowed by the sea.)
>
> (My translation)

For Sebald, this declensionism in the natural world is an apt setting for the pattern of decline he sees in human history. John Beck notes that "erosion is the subject of Sebald's *The Rings of Saturn*.... Along the Suffolk coastline the text finds everywhere the ruins of a great power" (Beck 2014, 75) and it is this visible decay that links the Suffolk landscape with the global stories recounted by the narrator.

One notable example of the decline of human civilization (for want of a better word) that is set in and mirrored by this desolate and marginal landscape is that of Orford Ness. This long spit of land off the Suffolk coast near Aldeburgh was the site of a military testing area during World War II and subsequent decades and was only abandoned (and therefore rendered accessible for non-authorized human visitors) in the early 1970s.[4] Sebald recounts his visit to Orford Ness in the eighth chapter of *The Rings of Saturn*.

> ... nach Norden und Süden glänzte das von einem mageren Rinnsal durchzogene Schlammbett des toten Flußarms, und voraus war nichts als Zerstörung.... Je näher ich aber den Ruinen kam, desto mehr ... wähnte ich mich unter den Überresten unserer eigenen, in einer zukünftigen Katastrophe zugrundegegangenen Zivilisation.
>
> (... to the north and south, in flashes of silver, gleamed the muddy bed of a dead arm of the river ... ahead lay nothing but destruction.... But the closer I came to these ruins ... the more I imagined myself amidst the remains of our own civilization after its extinction in some future catastrophe.)
>
> (281–2; 235, 237)

Sebald's narrator sees the signs of abandonment and decay and projects himself into the future, when the trajectory of decline is completed and human life and culture are extinct. Pippa Marland sees in the assemblage of defunct machinery and abandoned buildings a subtle reference to the Holocaust, writing that:

> On my first reading of the Orford Ness passage, the "primitive contraptions and fittings inside the bunkers, the iron rails under the ceilings, the hooks on the still partially tiled walls, the showerheads the size of plates, the ramps and soakaways" inexorably conjured images of the Holocaust in my mind, just as I believe they did for Sebald.
>
> (Marland 2014, 134)[5]

The Holocaust, only ever approached obliquely and in small details in Sebald's writing, seems to function as a cipher for the ultimate destruction, the ultimate failure, of the human project. Marland's exploration of the "affective power of material objects" (134) in her reading of this passage underlines the extent to which Sebald's descriptions of physical decay work emotively to invoke horror and shame. The post-Cold War wasteland of Orford Ness functions in Sebald's text as the culmination of all that is lost on England's eastern margin and in the history of humankind, to the extent that the narrator is temporarily displaced in time, imagining himself far in the future, looking back at the ruins of an unknown civilization.

What does Sebald's vista of decline mean in the context of the Anthropocene? (Bearing in mind, of course, that the term "Anthropocene" was first coined in 2000, two years after the publication in German of *The Rings of Saturn*.) The parallels that Sebald's text draws between ecological and anthropogenic processes, between human conquests of nature and conquests of peoples, suggest that he understands human history ecocentrically, that is, grounded in a context that is other-than-human. Human behaviour—from prehistoric land clearance and deforestation, to ocean pollution, to the fishing of herring—is deleterious to the wider environment, and ultimately this cycle of destruction returns to haunt the humans themselves. This is evident in the passage about deforestation that heads up Chapter VII of *The Rings of Saturn*:

> Die Entstehungsgeschichte dieser traurigen Gegend ist eng verbunden nicht nur mit der Bodenbeschaffenheit und den Einflüssen des ozeanischen Klimas, sondern, in weit entscheidenderem Maße, mit der über viele Jahrhunderte, ja über Millennien fortschreitenden Zurückdrängung und Zerstörung der dichten Wälder, die nach der letzten Eiszeit sich ausgebreitet haben über das gesamte Gebiet der britischen Inseln. In Norfolk und Suffolk sind es hauptsächlich Eichen und Ulmen gewesen, die über die Ebenen und in ununterbrochenen Wellen über die leichten Anhöhen und durch die Senken gingen bis hin an die Ufer des Meers. Die rückläufige Entwicklung setzte erst ein mit dem Auftauchen der ersten Siedler, die an den regenarmen östlichen Küstenstrichen, wo sie sich niederlassen wollten, Feuerbrände legten.... Was in Europa in der Vorzeit vom Feuer verschont blieb, wurde später geschlagen für den Haus- und Schiffsbau, sowie für die

> Gewinnung der für die Eisenschmelzen in ungeheuren Mengen benötigten Holzkohle.
>
> (The history of how that melancholy region came to be is closely connected not only with the nature of the soil and the influence of a maritime climate but also, far more decisively, with the steady and advancing destruction, over a period of many centuries and indeed millennia, of the dense forests that extended over the entire British Isles after the last Ice Age. In Norfolk and Suffolk, it was chiefly oaks and elms that grew on the flatlands, spreading in unbroken waves across the gently undulating country right down to the coast. This phase of evolution was halted when the first settlers burnt off the forests along those drier sections of the eastern coast where the light soil could be tilled.... Whatever was spared by the flames in prehistoric Europe was later felled for construction and ship-building, and to make the charcoal which the smelting of iron required in vast quantities.)
>
> (201–2; 169–70)

This description of human disregard for local ecosystems and hunger for fuel as a driver of deforestation and environmental degradation (and, ultimately, through the burning of coal, climate change) is aligned with the narratives of many environmental historians and advocates of the Anthropocene. Dipesh Chakrabarty (2009, 208) sees in the burning of fossil fuels and increased hunger for energy during the Enlightenment the beginnings of human geological agency ("the mansion of modern freedom stands on an ever-expanding base of fossil-fuel use"); Fredrik Jonsson sees the history of energy as key to understanding the genesis of the Anthropocene, writing that "energy history might reveal an unexamined basis for ideology, politics, and culture after the Enlightenment" (Jonsson 2015, 60).

All this speaks to *The Rings of Saturn* being a text that understands the role humans have played in shaping the planet around them on a large scale. The fact that Sebald's human histories are set for the most part during the period of the eighteenth, nineteenth, and twentieth centuries—during the Enlightenment, in Western terms—means that they coincide with the period of industrialization and scientific understanding that, in the view of Paul Crutzen, Dipesh Chakrabarty, Fredrik Jonsson, and many others, allowed humans to gain geological agency, and thus launched the Anthropocene. Can Sebald's chronicle of the decline of the human project and the destruction of the environment be read as a history of the makings of the Anthropocene?

To attempt such a reading, Sebald's view of the Enlightenment needs further examination. As we have seen in Chapter 2 of this book, the concept of the Anthropocene fundamentally calls into question the history of the Enlightenment as a history of the human quest for progress through science and understanding. The ambivalent status of the new dike in Theodor Storm's *Der*

Schimmelreiter seems to underpin Adorno and Horkheimer's arguments in *Dialektik der Aufklärung*, which certainly influenced Sebald (Martin 2013). Greg Bond notes that "Sebald is a modernist melancholic: his works are based on a critique of the history of progress—and enlightenment—that cannot conceive of a world in which progress is no longer a category" (2014, 42).

In other words, according to Greg Bond, Sebald's narrative of decline is dependent on the narrative of progress. If Sebald were to refute the idea of linear progress per se, then his chronicle of its decline and failure would be rendered void. Sebald uses the tools of Enlightenment education to subvert the basis for it. He illustrates his prose with images that often undermine or distort the meaning of the words; for example, the picture of the fish that accompanies the passages about the herring is, in fact, not a herring at all (it's a cod). Sebald illustrates falsification, rather than science; it is a false enlightenment, rather than no enlightenment at all. To return to John Beck:

> As a kind of counter-enlightenment joke, the book piles up the "facts"—the biographies of eminent figures, local history, natural history, arcane industrial techniques, international relations, imperial adventures—only to demonstrate that the increasing volume of interconnected information does not build towards an organized structure of control but to further instability and disappointment.
>
> (2014, 82)

This can be seen in the long passage about silk farming. The value of silkworms as objects of educative value is emphasized, as here, when Sebald quotes the Bavarian Privy Counsellor Joseph von Hazzi's 1826 book on silk production in Germany:

> Zudem sei die Beobachtung des unscheinbaren Insekts, wie es unter menschlichen Pflege sich stufenweise entwickelt und zuletzt die zartesten und nützlichsten Stoffe hervorbringt, ein höchst schickliches Mittel zur Bildung der Jugend.
>
> (In addition, observing this apparently insignificant insect, and how it develops in stages under man's care and brings forth the softest and most delicate of tissues, would provide an invaluable and fitting aid for the education of children.)
>
> (343; 290–1)

The process of harvesting silk from silkworms is presented as a trajectory of Enlightenment learning—from the monks who smuggled silkworms out of China, to the natural history of silkworms, their patterns of feeding, transmigration, and reproduction, to individual and state-led attempts to industrialize the production of silk for national profit. But this trajectory of ever greater understanding and better management is called into question when silk farming is taken up by the National Socialists in the 1930s. The object lesson of the silkworm is not just useful for the transmitting and understanding of natural history, but moreover:

> Bau und Besonderheiten des Insektenkörpers seien an ihr aufzuzeigen, desgleichen Domestikationserscheinungen, Verlustmutationen sowie die in der menschlichen Zuchtarbeit notwendigen Grundmaßnahmen der Leistungskontrolle, Auslese und Ausmerzung zur Vermeidung rassischer Entartung.
>
> (They could be used to illustrate the structure and distinctive features of insect anatomy, insect domestication, retrogressive mutations, and the essential measures which are taken by breeders to monitor productivity and selection, including extermination to preempt racial degeneration.)

Sebald has raised the spectre of Nazi eugenics and ethnic cleansing, and goes on to comment on the process by which the silkworms, having completed the process for which they have been selectively bred, are dispatched:

> … die Abtötung … über einem ständig am Sieden gehaltenen eingemauerten Waschkessel. Drei Stunden müssen die im flachen Körben ausgebreiteten Kokons über dem aus dem Schaff aufsteigenden Wasserdampf liegenbleiben, und wenn man mit einer Menge fertig ist, so fährt man mit der nächsten fort, so lange, bis das ganze Tötungsgeschäft vollendet ist.
>
> (… the killing, accomplished in this case … by suspending them over a boiling cauldron. The cocoons, spread out on shallow baskets, have to be kept in the rising steam for upwards of three hours, and when a batch is done, it is the next one's turn, and so on until the entire killing business is completed.)
>
> (348; 294)

Sebald's appropriation of eugenicist language here and his use of compound nouns "Zuchtarbeit," "Verlustmutationen," and "Tötungsgeschäft" evoke the bureaucratic and evasive language of Nazi officialdom and the Holocaust (in Hulse's English translations these terms are given as "the measures taken by breeders," "retrogressive mutations," and "killing business," inevitably losing some of this chilling association with the specific vocabulary of Nazi bureaucracy). His collection of knowledge, presented as a progression, a trajectory, is a product of the Enlightenment and simultaneously a subversion of the ideals of human progress and emancipation. The twin narratives of human achievement and human failure are deeply entangled; it is left to the reader to decide whether the bounty of herring catches in the early twentieth century represented human ingenuity or human cruelty, whether Joseph Conrad's *Heart of Darkness* calls to mind literary greatness or colonial depravity, or whether the rewilding of Orford Ness represents progress or decline.

Seeing Sebald's history of Enlightenment humanity in the context of Anthropocene thinking raises another question, that of the category of the human. One of the difficulties raised by Dipesh Chakrabarty is the impossibility of humans being able to comprehend themselves totally on the level of a species. On the topic of imminent global warming, he notes that:

> the ensuing crisis for humans is not understandable unless one works out the consequences of that warming. The consequences make sense only if we think of humans as a form of life and look on human history as part of the history of life on this planet. For, ultimately, what the warming of the planet threatens is not the geological planet itself but the very conditions, both biological and geological, on which the survival of human life as developed in the Holocene period depends.
>
> (Chakrabarty 2009, 211)

Arguably, Sebald's ecocentric approach to human history *does* at least attempt to contextualize it in the history of life on this planet, and thus prepares the way for not just the cultural, but the biological decline of humanity. As we have seen in his passages on the herring and the silkworms, the narrator of *The Rings of Saturn* draws parallels between the suffering human and non-human beings; most tellingly, perhaps, in the juxtaposition of the images of the millions of dead fish in Lowestoft and the piles of dead bodies in the concentration camp of Bergen-Belsen. This juxtaposition serves both to elevate the fish to potentially sentient, if unknowable, fellow creatures, and to relegate humanity to one amongst many species to suffer and inflict cruelty.

Another example of a biological or evolutionary interpretation of human behaviour is Sebald's remark about the phenomenon of "westing":

> Der aussichtslose Kampf wurde aufgegeben, man kehrt dem Meer den Rücken und baute, wann immer es die schwindenden Mittel erlaubten, nach Westen zu, ein zerdehntes, über Generationen sich hinziehendes Fluchtunternehmen, mit dem die langsam sterbende Stadt—reflexartig, könnte man sagen—eine der Grundbewegungen des menschlichen Lebens auf der Erde beschrieb. Auffällig viele unserer Ansiedlungen sind ausgerichtet und verschieben sich, wo die Verhältnisse es erlauben, nach Westen. Der Osten ist gleichbedeutend mit Aussichtslosigkeit.
>
> (They abandoned their hopeless struggle, turned their backs on the sea, and, whenever their declining means allowed it, built to the westward in a protracted flight that went on for generations; the slowly dying town thus followed—by reflex, one might say—one of the fundamental patterns of human behaviour. A strikingly large number of our settlements are oriented to the west and, where circumstances permit, relocate in a westward direction. The east stands for lost causes.)
>
> (191; 158–9)

Sebald is describing a genuine biological phenomenon. The nature writer Richard Mabey, in his book on East Anglia, observes the widespread tendency for fields, dwellings, and trackways to be "tilted towards the west," and cites Thoreau as a further advocate of the theory of westing:

> So was some fundamental fact of geography at the root of it, maybe one of the periodic slight falls or rises in the level of East Anglia relative to the sea? Or perhaps the prevailing east wind, or the transit of the sun, which made field-workers all "tack" one way? Were they, in prehistoric times, more sensitive than we are to the earth's magnetic field? … Or was Thoreau right in believing that "westing" is a fundamental urge in living things?
> (Mabey 2005, 81–2)

Sebald's reference to what appears to be an overriding instinct across the human species (and potentially other species) is a reminder to the reader of the biological nature of humanity. But this is also linked to the march of colonialism; Bernhard Malkmus (2015–16, 192) writes "in einer atemberaubenden Umkehrung der judeo-christlichen Erlösungsgeschichte beschreibt Sebald den frühen englischen kolonialismus als eine permanente Westverschiebung über zwei Kontinente hinweg" ("in a breathtaking reversal of the Judaeo-Christian story of salvation, Sebald describes early English colonialism as a permanent relocation westwards across two continents"—my translation). The pursuit of intellectual and conquestorial gain in the Enlightenment is aligned with an inherent biological urge to migrate westwards, towards the setting sun.

At this point in the text, following the mention of westing, with its bird's-eye view of humanity as a species, we shift to the biographical portrait of the romantic poet Charles Algernon Swinburne that immediately follows it. The juxtaposition of perspectives in *The Rings of Saturn* is a constant feature of the text; it is this, perhaps, that allows Sebald to come close to unpicking Chakrabarty's dilemma of understanding humanity as a species whilst not losing the distinctions between human groups (class, gender, colonizer/subaltern, industrialist polluter/pastoralist, etc.), and "putting global histories of capital in conversation with the species history of humans" (Chakrabarty 2009, 212). Sebald's technique is one of constantly zooming in and out between different perspectives, in spatial and temporal terms (from one spot in Southwold, to a trip via The Hague to Amsterdam, to Serbia during the reign of terror of the Ustacha secret police), as well as between different views of the human.

Sebald plays on the idea of perspective at many points in his book. *The Rings of Saturn* opens with an image of a window, with a view of empty sky interrupted by the black netting that is drawn over it. The window suggests a view that is defined above all by its limitation; the person looking out of that window has a severely circumscribed perspective on the world. Throughout the text, the narrator offers an array of different perspectives, each of which gives an imperfect view of the whole. From an aeroplane flying over the Netherlands, he notes that from that altitude, "eine über Jahrhunderte sich hinziehende Regulierungs-,

Kultivierungs- und Bautätigkeit hatte die gesamte Fläche verwandelt in ein geometrisches Muster" ("Over the centuries the land had been regulated, cultivated and built on until the whole region was transformed into a geometrical pattern") (112; 90). His preoccupation with the Rembrandt painting "The Anatomy Lesson of Dr. Nicolaes Tulp" is concerned with its deliberate distortion of perspective (22–7; 12–17). On viewing the memorial to the battlefield of Waterloo, a cupola with a painted panorama of the battle on its inner walls, the narrator writes:

> Das also, denkt man, indem man langsam im Kreis geht, ist die Kunst der Repräsentation der Geschichte. Sie beruht auf einer Fälschung der Perspektive. Wir, die Überlebenden, sehen alles von oben herunter, sehen alles zugleich und wissen dennoch nicht, wie es war.
>
> (This then, I thought, as I looked round about me, is the representation of history. It requires a falsification of perspective. We, the survivors, see everything from above, see everything at once, and still we do not know how it was.)
>
> (151–2; 125)

Sebald does not attempt to come to a complete depiction of human history, dwelling instead on the impossibility of any one view being anything other than a falsification of perspective. Through the combining of different perspectives, though, he comes perhaps closest to Chakrabarty's idea of a "conversation" between different historical scales, and thereby has planted his text at the heart of the Anthropocene debate. The retelling of the history of the Enlightenment as the history of the Anthropocene is no more contradictory, in light of Sebald's work, than any other ways of telling history; each of them provides a narrow, limited, and distorted perspective on the past.

So Sebald's *The Rings of Saturn* can be discussed as a literary account of the Anthropocene in three ways—first, through its ecocentric embedding of Enlightenment history in a planetary context; second, through its treatment of anthropogenic damage to the environment, besides natural processes of decay; and third, through its attempt to place humanity as a species into a historical category alongside more familiar ones, through multiple and conflicting perspectives. But given the narrator's melancholic fixation with the past and repeated lamentations of decline and destruction, it would perhaps be just as apt to describe this work as a chronicle of the dying days of the Holocene as an account of the beginnings of the Anthropocene. At such close hand, the seam between these two geological eras is impossible to discern. As Jonsson observes, "that dichotomy between deep time and civilization set the stage for one of the central claims of Anthropocene science: the recognition that human civilization has thrived only in the long summer of the Holocene" (2015, 63). *The Rings of Saturn* is a portrait, melancholic and nostalgic by turns, of all that was good and bad on the seam between the Anthropocene and the geological era which may prove to be, in biological terms, the only true age of humans.

Notes

1 Robert MacFarlane called *The Rings of Saturn* a deeply political travelogue on foot; in the same interview, Rebecca Solnit referred to it as "a Houdini book that escaped everything." Podcast interview with Robert MacFarlane and Rebecca Solnit, *Orion Magazine*, 3 January 2014. https://orionmagazine.org/article/live-event-robert-macfarlane-and-rebecca-solnit-on-nature-writing/.

2 Barbara Hui created her LITMAP project (http://barbarahui.net/litmap/) on W.G. Sebald's *The Rings of Saturn* in 2008—09 as part of her doctoral dissertation; this project is featured in Grant Gee's 2012 film *Patience: After Sebald*. Mandana Covindassamy's monograph *W.G. Sebald: Cartographie d'une écriture en déplacement* (Paris: Presses de l'université Paris-Sorbonne, 2014) was the result of her doctoral research.

3 All direct quotations from *The Rings of Saturn* in German are from Sebald (1997); all direct quotations in English are from the translation by Michael Hulse (Sebald, 1998).

4 The spit is under the management of the National Trust and is now known as Orford Ness National Nature Reserve, with limited public access provided by National Trust boats on designated days.

5 Pippa Marland quotes Sebald from the English edition. The German passage reads "… die primitiven Anlagen im Innern der Bunker, die Eisenbahnschienen unter den Decken, die Haken an den zum Teil noch gekachelten Wänden, die tellergroßen Brausen, die Rampen und Sickergruben …" Sebald, *Ringe des Saturn*, 282.

References

Agnew, John A. 2011. "Space and Place." In *The Sage Handbook of Geographical Knowledge*, edited by John A. Agnew and David N. Livingstone, 316–30. London: Sage.

Alexander, Neal. 2015. "On Literary Geography." *Literary Geographies* 1 (1): 3–6.

Augé, Marc. 1995. *Non-Places: Introduction to an Anthropology of Supermodernity*. Translated by John Howe. London: Verso.

Beck, John. 2014. "Reading Room: Erosion and Sedimentation in Sebald's Suffolk." In *W.G. Sebald—A Critical Companion*, edited by J.J. Long and Anne Whitehead, 75–88. Edinburgh: Edinburgh University Press.

Bond, Greg. 2014. "On the Misery of Nature and the Nature of Misery: W.G. Sebald's Landscapes." In *W.G. Sebald—A Critical Companion*, edited by J.J. Long and Anne Whitehead, 31–44. Edinburgh: Edinburgh University Press.

British Department for Environment, Farming and Rural Affairs. 2010. "Waveney Coastal Change Pathfinder." http://webarchive.nationalarchives.gov.uk/20130403070903/http://archive.defra.gov.uk/environment/flooding/manage/pathfinder/waveney.htm (last accessed 3 July 2018).

Catling, Jo. 2008. "W.G. Sebald: Ein 'England-Deutscher?' Identität—Topographie—Intertextualität." In *W.G. Sebald: Intertextualität und Topographie*, edited by Irene Heidelberger-Leonard and Mireille Tabah, 25–53. Münster: LIT.

Chakrabarty, Dipesh. 2009. "The Climate of History: Four Theses." *Critical Enquiry* 35 (2): 197–222.

Clark, Timothy. 2011. *The Cambridge Introduction to Literature and the Environment*. Cambridge: Cambridge University Press.

Cook, Jon, ed. 2014. *After Sebald: Essays and Illuminations*. Woodbridge: Full Circle.

Covindassamy, Mandana. 2014. *W.G. Sebald: Cartographie d'une écriture en déplacement*. Paris: Presses Universitaires de Paris Sorbonne.

Crutzen, Paul J. and Eugene F. Stoermer. 2000. "The Anthropocene." *IGBP Newsletter* 41: 17–18.

Gee, Grant. 2012. *Patience: After Sebald*. Film. Soda Pictures UK.

Goodbody, Axel. 2010. "Die Ringe des Saturn und Solar: Sinnbilder und Schreibstrategien in literarischen Stellungnahmen zur ökologischen Krise von W.G. Sebald und Ian McEwan." In *Ökologische Transformationen und literarische Repräsentationen*, edited by Maren Ermisch, Ulrike Ruse, and Urte Stobbe, 131–48. Göttingen: Universitätsverlag Göttingen.

Goodbody, Axel. 2014a. "Editorial." *Ecozon@* 5 (2): i–iii.

Goodbody, Axel. 2014b. "Ecocritical Theory: Romantic Roots and Impulses from Twentieth-Century European Thinkers." In *Literature and the Environment*, edited by Louise Westling, 61–74. Cambridge: Cambridge University Press.

Heise, Ursula. 2008. *Sense of Place and Sense of Planet: The Environmental Imagination of the Global*. Oxford: Oxford University Press.

Hui, Barbara. 2008–09. "Litmap (beta): Book Title: The Rings of Saturn: An English Pilgrimage [Die Ringe des Saturn: Eine englische Wallfahrt]. Author: W.G. Sebald (trans. Michael Hulse)." http://barbarahui.net/litmap/ (last accessed 3 July 2018).

Jonsson, Fredrik Albritton. 2015. "Anthropocene Blues: Abundance, Energy, Limits." In *The Imagination of Limits: Exploring Scarcity and Abundance*, edited by Frederike Felcht and Katie Ritson. *RCC Perspectives* no. 2, 55–63.

Mabey, Richard. 2005. *Nature Cure*. London: Chatto & Windus.

Macfarlane, Robert and Rebecca Solnit. 2014. Podcast interview for Orion Magazine, 3 January 2014. https://orionmagazine.org/article/live-event-robert-macfarlane-and-rebecca-solnit-on-nature-writing/ (last modified 3 January 2014).

Malkmus, Bernhard. 2015–16. "Naturgeschichten vom Fisch, oder: Die Angst vor dem Anthropozän." *Scheidewege: Jahresschrift für skeptisches Denken* 45: 183–200.

Malkmus, Bernhard. 2017. "Diffuse Dwelling: Landscape in Graham Swift and W.G. Sebald." In *Readings in the Anthropocene: The Environmental Humanities, German Studies, and Beyond*, edited by Sabine Wilke, 263–95. Rochester NY: Bloomsbury.

Marland, Pippa. 2014. "'Heaps of Scrap Metal and Defunct Machinery': Assemblages, Ethics, and Affect in W.G. Sebald's Orford Ness." *Ecozon@* 5 (2): 123–39.

Martin, James R. 2013. "On Misunderstanding W.G. Sebald." *Cambridge Literary Review* 4 (7): 123–38.

McCrum, Robert. 2013. "W.G. Sebald's Quietly Potent Legacy." *Guardian*, 13 May 2013. www.theguardian.com/books/booksblog/2013/may/13/wg-sebald-legacy (last modified 13 May 2013).

Mønster, Louise. 2009. "At finde sted: En introduktion til stedbegrebet og dets litterære potentiale." *Edda* 4, 357–73.

Riordan, Colin. 2014. "Ecocentrism in Sebald's *After Nature*." In *W.G. Sebald—A Critical Companion*, edited by J.J. Long and Anne Whitehead, 45–57. Edinburgh: Edinburgh University Press.

Sebald, W.G. 1997. *Die Ringe des Saturn: Eine englische Wallfahrt*. Frankfurt am Main: Fischer.

Sebald, W.G. 1998. *The Rings of Saturn*. Translated by Michael Hulse. New York: New Directions.

Weston, Daniel. 2011. "The Spatial Supplement: Landscape and Perspective in W.G. Sebald's The Rings of Saturn." *Cultural Geographies* 18 (2): 171–86.

Wood, James, Will Self, Iain Sinclair, and Robert Macfarlane. 2013. "W.G. Sebald: Reveries of a Solitary Walker." *Guardian*, 20 April 2013. www.theguardian.com/books/2013/apr/20/wg-sebald-reveries-solitary-walker (last modified 20 April 2013).

4 Landscape as palimpsest

East Anglia in British "new nature writing"

W.G. Sebald's well-trodden route through East Anglia embeds the landscape in a global story of simultaneous progress and decline. This chapter also takes as its focus East Anglia as a site for the imagination of environmental change and environmental crisis through its exploration of British "new nature writing." This is a mode of writing which, while eschewing the formal "strait-jacket" of genre,[1] nonetheless finds coherence in its use of literary techniques to advance a political and ethical standpoint with relation to the natural environment. Most of the texts in this book take the form of fictional writing and none of them have been marketed explicitly as an environmentally or ecologically oriented work: their use of environment goes beyond mere setting, and has an impact on plot and characters (see Lawrence Buell's criteria for an environmental text in the Introduction), but the novels and short stories do not have an overtly expressed political agenda. New nature writing has been posited as an ethically charged response to the environmental crisis of the Anthropocene, and the fragile landscapes of the English east coast figure strongly within the genre.

There is a general consensus that nature writing is experiencing a renaissance in Great Britain. This fact is reflected in the large volumes of titles marketed under this genre label, but also its high visibility. A new award created in 2014, the Thwaites Wainwright Prize, seeks "to reward the best writing on the general outdoors, nature and UK-based travel writing,"[2] but nature writers have also been well represented in general literary awards. Richard Mabey won the Whitbread Biography of the Year for his life of the celebrated naturalist Gilbert White; Robert Macfarlane has won numerous prizes for his work, including the Guardian First Book Award and the Sunday Times Young Writer of the Year Award (both awarded in 2003 for his book *Mountains of the Mind*). Macfarlane was also shortlisted for the prestigious Samuel Johnson Prize for Non-Fiction for *The Old Ways* in 2012. More recently, in 2014, the Samuel Johnson Prize was won by Helen Macdonald for her book *H is for Hawk*, a meditation on her relationship with a goshawk.

In his introduction to the *Granta* special issue entitled "The New Nature Writing," Jason Cowley attributes the revival of nature writing to the threats to the ecological system of the planet:

> Few would doubt that we are living at a time of emergency. The world's population ... is projected to rise to 8.5 billion by 2030. It is understood now just how quickly the earth is warming, because of the increase in the atmospheric concentration of greenhouse gases arising from human activity. If the earth continues to warm at its present rate, we know what our fate will be, and yet we seem intent on destroying ourselves.... One need not be a pessimist to predict some kind of Malthusian denouement to the human story if we are unable or unwilling to alter our ways of being.
>
> (Cowley 2008, 9)

The sentiment is echoed by Robert Macfarlane, one of the best-known exponents of British nature writing:

> It is no coincidence that a literature celebrating the natural world should have emerged at a time when the natural world is so conspicuously under threat. The past 15 years have seen the Deepwater Horizon blowout in the Gulf of Mexico, the break-up of the Antarctic ice shelf, widespread habitat destruction, further evidence that we are living through the sixth great extinction pulse, and the slow-motion emergency of climate change. British nature writing is energized by this sense of menace and hazard.
>
> (Macfarlane 2013b, 167)

Both Macfarlane and Cowley make clear the ethical concerns that underpin the revitalized interest in writing about nature. New nature writers produce their texts against the backdrop of the overwhelming sense of urgency caused by the manifold threats to the environment, whether their writing takes the form of awed examinations of the intricacy of ecologies, elegies for what is, or is about to be lost, or reflections on the entangled natures that arise from human activity in the landscape. This sense of menace is particularly acute on England's low-lying eastern seaboard, where coastal erosion is being accelerated by rising sea-water levels and the drying out and shrinking of the drained land. Some of this landscape is lost to the sea each year, and it is at the forefront of conservationists' discussions about managed retreat and rewilding (Rotherham 2013; in the mainstream media, see e.g., McVeigh 2015; Barkham 2014). It is thus not surprising to find that East Anglia is well represented in recent works of nature writing, by authors such as Robert MacFarlane (*The Wild Places* 2007 and *The Old Ways* 2013a), Richard Mabey (*Nature Cure* 2005), Jules Pretty (*This Luminous Coast* 2012), James Canton (*Out of Essex: Re-Imagining a Literary Landscape* 2013), Helen Macdonald (*H is for Hawk* 2014), and Mark Cocker (*Claxton: Field Notes from a Small Planet* 2014). In this chapter, I will look in particular at two texts by Cambridge-based Robert Macfarlane.

English-language nature writing has a long tradition on both sides of the Atlantic; in recent decades, it has been a mainstay subject of the ecocritical scholarship coming out of North America. American nature writing has long enjoyed more prestige than its British counterpart; the variety of landscapes on the American continent, the rich imaginary provided by the nineteenth-century idea of "wilderness" as advocated by John Muir and the Sierra Club, and latterly the drive to reflect a landscape whose indigenous inhabitants are marginalized or gone, have all been strong currents in American texts. Besides groundbreaking ecological writers Rachel Carson and Aldo Leopold, and influential contemporary writers such as Barry Lopez, Annie Dillard, and Rebecca Solnit, Henry David Thoreau's *Walden* has been a seminal text for the field of ecocriticism and continues to serve as a touchstone for contemporary American writing and ecocritical scholarship (with Lawrence Buell's 1995 book *The Environmental Imagination: Thoreau, Nature Writing, and the Formation of American Culture* a particularly influential work). American nature writing has greatly influenced the British tradition in recent years, but the two nonetheless remain very different in both their origins and their current trajectories.

Jason Cowley explains the contrast between "old" nature writing and the newer phenomenon to which the 2008 *Granta* volume is devoted: "We were interested less in the old nature writing—by which I mean the lyrical pastoral tradition of the romantic wanderer—than by writers who approached their subject in heterodox and experimental ways" (Cowley 2008, 10). This "lyrical tradition," embodied in English literature by characters such as John Clare or William Wordsworth, has nature as a refuge from an encroaching modernity. In contemporary writing of the Anthropocene, no such distinction between nature and modernity is possible: anthropogenic developments, such as the electric lights that pollute the night sky, the microplastic waste that is poisoning filter feeders such as whales, and the carbon dioxide that is warming the atmosphere are examples of the ubiquity and porosity of human influence. Indeed, the agency of humans is called into question, since we now "inhabit a dimension crisscrossed by vibrant forces that hybridize human and nonhuman matters," as Serenella Iovino and Serpil Oppermann put it in the introduction to their book *Material Ecocriticism* (Iovino and Oppermann 2014, 5). The Enlightenment tradition of seeing humans as separate from nature has been replaced by the growing awareness of a new and troubling kind of interdependence, and a new kind of nature writing is the response. Cowley goes on to assert that:

> the best new nature writing is also an experiment in forms: the field report, the essay, the memoir, the travelogue. If travel writing can often seem like a debased and exhausted genre, nature writing is its opposite: something urgent, vital and alert to the defining particulars of our times.
>
> (Cowley 2008, 10)

Robert Macfarlane describes nature writing thus:

> It is widely agreed that Britain is going through a golden age of nature writing, but no one seems sure quite how to define it. Ragtag, wayward and polymorphous, it folds in aspects of memoir, travel, ecology, botany, zoology, topography, geology, folklore, literary criticism, psychogeography,

anthropology, conservation and even fiction. Most distinctive, to my mind, is its tonal mix of the poetic and the scientific and analytical.

(Macfarlane 2013b, 166)

Macfarlane has made clear that he sees W.G. Sebald as one of the most important influences on the new nature writing, in particular his rejection of the conventions of genre. Sebald's *The Rings of Saturn* is a hybrid text, and echoing this, Macfarlane's definition of nature writing is a hybrid one. It draws upon a wide range of textual forms, including many that are not typically literary (such as geological surveys and field notes) and can combine fictional and non-fictional elements. This medley of styles indicates an understanding of the human and non-human environment that is complex and multi-layered, and that goes beyond a purely aesthetic response, inscribing scientific knowledge and information into its narrative. Tim Clark writes that the form of the literary essay that is often adopted for nature writing "suits the often perplexingly interdisciplinary nature of environmental issues. Its freedom can embrace material from diverse sources that would not be admitted in a scientific paper or a piece of historical research" (Clark 2011, 36).

The juxtaposition of different textual forms creates a narrative that oscillates between objective and subjective viewpoints of the landscape, between peer-reviewed sources and personal anecdotes. In *Nature Cure*, Richard Mabey interlaces his own observations and dreams about the landscape with his naturalist's knowledge of its history and management.

> Flint is one of the most widely used raw materials on earth, yet no one is sure how it is formed, except that it is a kind of silica, leached out of chalk, and probably metamorphosed in the heat and pressure of volcanoes … I once found an eolith, one of those cryptic "dawnstones" whose sharp edges maybe signs of those first shapings, or then again be entirely natural, a product of pack-ice jostling—geology posing as craft. It was like the last two joints of a finger, with a point at one end.... It's my heartstone.
>
> (Mabey 2005, 93)

The effect of this shifting of perspective is to reveal a relationship with the natural world that is both deeply informed by a geological and historical understanding of it, and that draws on subjective impression and meaning to aesthetic effect. This kind of oscillation, also reminiscent of Sebald's preoccupation with different physical perspectives, is a common feature of the new nature writing: all of the works mentioned in this chapter are recounted by an intradiegetic narrator who has a personal relationship with the wildlife and the landscapes through which she moves. Two of the works (*H is for Hawk* and *Nature Cure*) are explorations of the narrator's breakdown in mental health and subsequent recovery, a subjective and personal journey that shapes the narrative from beginning to end and colours its interactions with the natural world.

In *H is for Hawk*, the narrator's exploration of her grief following the sudden death of her father is as intrinsic to the narrative as her relationship with the goshawk Mabel and their hunting expeditions in the countryside around Cambridge. Helen Macdonald gropes towards an expression of an avian, hawk's-eye view of the world to complement the human-subjective and objective observations of nature in her book.

> Her eyes can follow the wingbeats of a bee as easily as ours follow the wingbeats of a bird. *What is she seeing?* I wonder, and my brain does backflips trying to imagine it, because I can't.... The hawk can see colours I cannot, right into the ultraviolet spectrum. She can see polarised light too, watch the thermals of warm air rise, roil and spill into clouds, and trace, too, the magnetic lines of force that stretch across the earth. The light falling into her deep black pupils is registered with such frightening precision that she can see with fierce clarity things I can't possibly resolve from the generalised blur. The claws on the toes of the house martins overhead. The veins on the wings of the white butterfly hunting its wavering course over the mustards at the end of the garden.
>
> (Macdonald 2014, 96)

The oscillating perspective creates a sense of a genre that is in movement in its search for a way of representing a subjective (and thus inevitably entirely anthropocentric) experience of the natural world that is nonetheless able to express a landscape in terms of its non-human parameters—its geology, non-human wildlife, ecosystems, and history. Explicitly and implicitly, new nature writers are drawing on technological advances that allow us to understand our environment from ever new vantage points and in new ways. The iconic "blue planet" image from the 1970s, often cited as a moment when the human perception of its own ecosystem shifted towards recognition of Earth's vulnerability, is only part of this; as Thomas Lekan has argued, many other influences on our view of the world, not least those of wartime pilots flying over destructive human populations (incidentally a theme in Helen Macdonald's work, and in W.G. Sebald's), have played their part in changing how we see the world around us (Lekan 2014).

Landscape as text

Seeing the landscape in objective terms often starts out from cartography. The forensic view of the English landscape as layers of history was advanced by William George Hoskins in his book *The Making of the English Landscape*, published in 1955. Hoskins taught his readers to see the traces of earlier settlement patterns, agricultural use, and historical decision-making in the modern environment, and is generally regarded to be the inventor of the academic discipline of landscape history in Great Britain; his evocative sense of landscape as palimpsest has also had a deep and abiding influence on local historians, nature lovers, and nature writers over the decades since his book's publication. Writing about Lincolnshire and East Anglia in the section entitled "Marsh, Fen and Moor," Hoskins notes:

> There are certain sheets of the one-inch Ordnance Survey maps which one can sit down and read like a book for an hour on end, with growing pleasure and imaginative excitement.... Such an exciting map is that of the country around the Wash, particularly the country on the western and southern sides—the Lincolnshire and Norfolk Marshland.... The very names like Bicker Haven and Fleet Haven, now many miles inland; or Moulton Seas End and Surfleet Seas End and Seadyke Farm, all far back today from the shore-line; the so-called Roman Bank twisting across the open levels, and the chapels far out on the salt marshes—all these matters of observation set the mind working at once about the past history of this piece of country.
>
> (1955, 95)

The description of the landscape as palimpsest is a feature common to a great deal of British nature writing. In his 2014 book centred around the village of Claxton in Norfolk, Mark Cocker, considered a leading exponent of the genre (Philip Hoare, writing in the *New Statesman* in 2015, asserted that Cocker's *Crow Country* "could be seen as the founding texts of 'new nature writing'"; Hoare 2015, 56), writes of a ruined Norman church, St. Saviour's:

> In the early modern period its tower collapsed. In the nineteenth century the roof vanished, and where the pews once stood rises a tangle of nettle, bramble and other faithful attendants of human indifference.... Wild rabbits, which would have been almost unknown in Britain when St. Saviour's was first built, have riddled the remains of the chancel wall with deep holes. In my imagination I follow the animals down their burrows and where they now scratch their fleas are the earliest stones ever laid at the church. I think of the medieval labourers who placed them there and I wonder what they might have done had I told them what is now established fact: that the black flint lumps in their hands, carefully slotted into the walls layer by layer, were once sea urchins and sponges living in the warm waters that stood over this spot about eighty million years ago.
>
> (Cocker 2014, 11)

The ruined church and the land it stands on serve as a space to imagine the history of the landscape; once again, the juxtaposition of timescales refract the Anthropocene collision of natural and human histories by "engaging dilemmas of scale," as environmental historian Emily Wakild has put it in her thought-provoking essay on the challenge of scale (Wakild 2014, 29). This small corner of the East Anglian landscape is a nexus in which different historical timescales—human endeavour over centuries, but also the lifespans and endeavours of rabbits and fleas, and the geological changes over millions of years—come together. Similar passages can be found in almost any other nature writer.

An awareness of the landscape's past is the foundation of understanding its present incarnation; and whereas Hoskins concentrated primarily on the human timescale and the human impact on the landscape, current nature writers also display their understanding of both the deep-time geological forces that have shaped the world around them, and also of the non-human agencies that have left traces on the land.

In East Anglia, a key feature of the landscape is its liminality, especially on the coast and in the sodden terrain of the Norfolk Broads and the Fens. The landscape here is often just as much a waterscape: it is historically unstable and the pervading wetness and flatness give it its particular character. In his *Nature Cure*, describing his moving (or "flitting") from the Chilterns to Norfolk, Richard Mabey writes, "I've seen enough of wet places to know they can be mercurial and unpredictable … the wet is older than the wood, but it is the domain of the present, and sometimes, it feels, of the future" (Mabey 2005, 11). He writes of the landscape's history:

> South and east of Norwich are the rivers Waveney, Bure and Yare. For thousands of years, as the level of the North Sea rose and fell, they periodically flooded the region, and turned much of it into a swamp. A great fen stretched all the way from the brackish marshland of the three rivers' common estuary … right up to the edge of the Breckland sands. A swamp of reed, lagoons, and sagging alder woods, seething with life, with spoonbills, ospreys, cranes and otters, and creatures like pelicans and beavers that are now just memories. Even the human inhabitants had to be amphibious. (9)

The palimpsest of the wet landscape is harder to read, and the imagination of the past of the landscape, brought to "seething" life through the text, shows the way in which the text both draws on physical features of the place and also serves to reanimate and recharge the imagined landscape across time. This visualization of the landscape in different temporal moments is a reason why these lowlands are such a rich site for the imagination of the Anthropocene. The layers of influence on the watery landscape are also put in the context of modern-day extreme weather events and climate change, here in a passage from *H is for Hawk* by Helen Macdonald:

> The skies broke and it rained and rained. The news was full of inundations and drowned cities; lost villages at the bottom of lakes, flash floods spilling over the M4 motorway to strand holiday traffic; kayaks on town streets in Berkshire; rising sea levels; the discovery that the English Channel was carved out by the bursting of a giant superlake millions of years ago. And the rain continued, burying the streets in half an inch of bubbling water, breaking shop canopies, making the River Cam a café-au-lait surge, thick with broken branches and sodden undergrowth. My city was apocalyptic. (2014, 15)

Here, the layering up is both visual and syntactical. In contrast to the vertical movement of the falling rain, the imagery is of layers of water, flattening everything as they rise. The waters cover ancient villages and the flat tarmac of the motorway, spilling over the edges; the sea is rising and a prehistoric lake burst its banks; the River Cam surges. The paratactical sentence construction adds to the sense of flood after flood, filling the land. The landscape of East Anglia here is a menacing waterscape, a wet palimpsest of rising tidemarks that make clear its significance in a world threatened by global warming and rising seawater levels. Both Mabey's and Macdonald's descriptions of the seeping wet recall myths of creation—both the biblical myth of the deluge that made the world anew, and the evolutionary narrative that has sentient life crawling out of the primeval swamp onto dry land. Water is dynamic, creating and transforming, and these two texts couple this dynamism to a current moment in which the waters are rising steadily. Our world is being remade, again; the palimpsests are textual and structural as well as visual.

And indeed, the idea of landscape as palimpsest is given double meaning in a new British nature writing that is a profoundly intertextual form of literature; the term palimpsest is here close to its original meaning of layers of earlier texts that show through to the surface of the manuscript. In *H is for Hawk*, the autobiographical story of the narrator's grief following the death of her father, and her reflections on the natural world through the care of her goshawk Mabel, are linked together and given narrative structure through an earlier text about human failure and the task of an austringer, T.H. White's *The Goshawk*, published in 1951. In Richard Mabey's *Nature Cure*, the chief intertext is the poetry of John Clare: like Macdonald's book, Mabey's is an account of his recovery from depression through his interactions with the environment and wildlife of East Anglia, and he draws parallels between himself and Clare, noting that they were even both treated in the same hospital for problems with their mental health, and both found solace in the natural world. Clare's grief at the loss of the commons during enclosure finds an echo in Mabey's writing at a time when the entire planet's ecology is in crisis. Other works of nature writing do not use a single intertext as a narrative device, but all draw widely on named and implicit textual sources of all kinds. The environments that provide the "nature" of nature writing are both constituted and mediated by literary and scientific text.

In a podcast interview with the American nature magazine *Orion*, Robert Macfarlane traces his genesis as a nature writer back to an encounter with a book:

> Whenever I talk to anyone who's actively involved in landscape or nature, and I ask them why they are, they almost always track it back to a place and to a book. And I can do the same, and the book is *Arctic Dreams* by Barry Lopez. And I came across that in the late nineties when I was travelling in Canada, and I was climbing and walking out on Vancouver Island and in the Rockies, and in a bookstore I came across a secondhand copy and it

> changed my life, it changed the direction of my thoughts, it provided many of the textures of my writing.
>
> (Macfarlane and Solnit 2014, 05:55 mins)

Macfarlane's starting point of a book is telling; experience of nature is mediated through text. Macfarlane also talks about the profound influence of W.G. Sebald on the new nature writing, an influence that is most noticeable in Macfarlane's own work:

> W.G. Sebald's *The Rings of Saturn*, which was written by obviously an emigré German who came to live in rural East Anglia near Norwich, and then, as we know created this great piebald, polymorphous form that looks a little like "traditional" (as it were) nature writing in that it's first-person-led, it mixes its forms, it's faintly antiquarian, quiet and recursive in its modes … it's a deeply political travelogue on foot.… It's about how history, and in some cases atrocity, soaks landscape.
>
> (17:40 mins)

Sebald's hybrid prose and its refuting of genre categories provides a template for new nature writing that is intertextual and critical in nature. The polyphony of attributed and unattributed intertexts and references in Sebald's work is also expressive of the idea of the landscape as a layering of palimpsests that inform and contradict one another. The intertexts that inform modern environmental prose range across genres and modes, making up what Macfarlane has referred to as "an ecology that includes the field note, the scientific journal, the polemic, the committee report" (42:01 mins).[3]

The intertextual nature of environmental prose texts thus also invites a multilayered understanding of the idea of "reading" a landscape. Naturalists and ecologists say they are reading a landscape when they put various signs together to make meaning. Mabey gives an example of this:

> We meandered down to the river, found otter footprints (all five toes to the front and known up here, somewhat confusingly, as "seals"). We saw posts where barn owls had rested, and a thin trail that wound into the reed-swamp and just vanished—as had the escaped heifer that made it.… Alongside the main track were what seemed to be groups of individual alder trees. But they were close together for their size, tilted outwards like sheaves in what, when you looked closely, were loose, hollow rings.… What looked like individual trees were part of a single plant. I guess that these alders were also old stools, and at 7 or 8 feet in diameter could well have been five hundred years old.
>
> (2005, 71)

Mabey is able to "read" the landscape, and, through his text, so are we. These examples show how the idea of palimpsest links the reading of the landscape and the reading of its literature. Both the genre of new nature writing and its subject are made up of layers of history and meaning that overlap and inform one another, and moreover, the two things interact with each other. Our understanding of landscape is built on textual palimpsests, and textual landscapes written through physical traces of older features. Understanding the way in which landscape is constituted and mediated in the human imagination is part of the nature writing project of re-engagement with the more-than-human environment.

After Sebald: Robert MacFarlane's "Ghost Species" (2008) and "Silt" (2012)

In this section I will look at two short pieces of writing by Robert Macfarlane that engage specifically with the landscapes of the North Sea lowlands in East Anglia; the first, "Ghost Species," with the Norfolk Fens, and the second, "Silt," with the Essex coast. Both were initially published in editions of *Granta: The Magazine of New Writing*, four years apart.

In "Ghost Species," the narrator travels to Norfolk to explore a liminal landscape that he considers to be almost a world of its own, and indeed, as he explains, was in effect a separate kingdom for a long period of its human history. Entering the Fens, he writes, is "like crossing a border into another world" (Macfarlane 2008a, 109). The reasons for this are to be found in both the relentless flatness of the land and its history, the stamp it bears of human intervention in the environment. It is rectilinear, flat, with sharp angles and clear horizons. "The Fens were once mostly water … brackish swamps and reed beds, interspersed with islands and causeways of raised but marshy ground" (111) Macfarlane writes, but then goes on to explain how the Fens were drained under Vermuyden, in a passage strongly reminiscent of Hoskins. "The Fens terraqueous past is still visible in contemporary maps, most obviously in the generic names common in the region: the Sluices and Bridges, the Lodes, Leams and Dykes, the Drains and the Mills" (112).

Macfarlane reflects on the significance of the Norfolk Fens in the video interview produced by *Granta* in conjunction with the special issue of the magazine in which "Ghost Species" appears:

> It's a lateral landscape and that's a hard … a much harder landscape to read. And now I'm starting to see it much more … and you know, places don't get much flatter than the Fens. You could fax them. The other thing about the Fens is that they're humanly constructed. That's what made them I think an interesting setting, particularly in the context of this issue of *Granta*. They were drained in the seventeenth century, they were really this sort of terraqueous land, half-sea, half-marsh, with a few causeways running between [it], and inhabited by people who'd adapted their ways of life to terraqueous terrain. So as soon as you step into the Fens, you're into a humanly generated place.
>
> (Macfarlane 2008b, 03:17 mins)

The Fens are presented as both difficult to read and important to write about; the "nature" of the Fens is also slippery, since they are not natural but constructed landscapes, and this fact makes them significant as a site for thinking about environmental change. "Ghost Species" is a piece in the tradition of Sebaldian melancholy, circling around what is being lost in and from the landscape. The title is a term taken from conservation science, where it is used to denote species that are close to extinction, living on beyond the ecological parameters that have sustained them over hundreds or thousands of years. Macfarlane contemplates this concept as he talks to Eric Wortley, a 98-year-old smallholder from the Fens who has scarcely left the surroundings of his farm since his birth there. The lifestyle Eric Wortley represents, a human way of living in the Fens that was virtually unchanged over centuries, is close to extinction; large-scale agriculture has overtaken small family farms, changing both the landscape of East Anglia and its inhabitants:

> Historically, the idea of ghost species has been confined to the non-human kingdoms. But sitting in Eric's kitchen that January day, it seemed clear that there were also human ghosts: types of place-faithful people who had been out-evolved by their environments—and whose future disappearance was almost assured.
>
> (Macfarlane 2008a, 126)

The Sebaldian preoccupation with loss is underlined by the use of images alongside the text, in this case provided by the photographer Justin Partyka, who is named in the text and accompanies the author on his trip to the Fens. The atmospheric photographs were taken to record a dying way of life for posterity, depicting the elderly Eric Wortley in front of his stove in the farmhouse he was born in, a fisherman with a catch of eel, and a smallholder on his field of sugar beets.

In his influential book *Flight Ways: Life and Loss at the Edge of Extinction*, the philosopher Thom van Dooren refers to the fate facing these ghost species as "the dull edge" of extinction:

> the edge of extinction is more often a "dull" one: a slow unravelling of intimately entangled ways of life that begins long before the death of the last individual and continues to ripple forward long afterward, drawing in living beings in a range of different ways.
>
> (van Dooren 2014, 12)

Van Dooren's book focuses on five different avian species threatened with extinction "woven into relationships with a diverse array of other species, including humans." He argues persuasively that "these relationships matter. This is true at the best of times, but in times like these, when so many species are slipping out of the world, these entanglements take on a new significance" (4). His book is an exhortation to combine the tools of storytelling with the knowledge of biologists and ethologists and conservation, "to insist on a truth that is not reducible to populations and data: a fleshier, more lively truth that in its telling might draw us all into a greater sense of accountability" (10). Van Dooren's manifesto for an ethical storytelling that moves the discussion about species loss from the domain of objective science to subjective emotional engagement is similar to Macfarlane's own approach in "Ghost Species." Macfarlane embraces the category slippage between human and animal, mourning the loss of both bird species and the "place-faithful" humans that lived in and with this place. He also writes human language into his ecology of place, implicitly also mourning the loss of the words ("fulfa" for thrush, "mabish" for mistle thrush; Macfarlane 2008a, 127) that once populated the Fens.

Van Dooren's book on extinction is focused on bird life. Many other new British nature writers have a particularly close relationship with birds. Mark Cocker and Richard Mabey are both naturalists and birdwatchers, having collaborated on the reference book *Birds Britannica*;[4] Helen Macdonald is an austringer and falconer, and her book is concerned both with wild birds and those that are captive, entangled for hundreds of years with the lives and ambitions of humans. Macfarlane too displays a keen knowledge of and interest in the bird life that surrounds him. The centrality of birds to new nature writing reflects their status as indicator species in many ecosystems. Mark Cocker writes that "birds often play the role of ambassador in our entire relationship with nature. For environmentalists they are collectively the miner's canary, their populations helping us to gauge the health of natural environments from the inner city to the remote Arctic tundra" (Cocker 2005, ix).

In "Ghost Species," the birds, like the traditional human inhabitants, are scarce. The narrator names the creatures that he sees or remembers in the landscape—the rooks, the migrating geese, the pair of deer—but, by and large, the landscape is empty, man-made, rectilinear. He and his companion see few people besides themselves and many signs of abandonment, such as the tabernacle or derelict barn, "whose roof was being prised off by fingers of ivy" and the children's chairs rooted to the spot with nettles (Macfarlane 2008a, 116). This emptiness and the wide open spaces of modern agriculture are contrasted with Eric Wortley's memories of the birds:

> Ah, well, nowadays you don't see hardly no animals on the land. A hare or two, mebbe. No birds, hardly any birds.... We used to have birds' eggses by the hundred.... Kestrels' eggs, sparrowhawks' eggs.... They were thicker then, the birds. Tomtits, blue tits, jenny wrens, especially the jenny wrens. I don't see the wrens any more.
>
> (127)

The birds have departed, or are ghosts of a past ecosystem, like Eric Wortley himself. But this lament of loss is countered at the very end of the piece by the sighting of a pair of little egrets (*Egretta garzetta*) rising into flight from a field as the narrator is on his way home. The symbolism of the moment is underlined by the simultaneous appearance of a rainbow in the dark sky, and by the dazzlingly white plumage of the bird itself: the little egret is not a species threatened with extinction, but a returner, a pioneer carving a new niche for itself.

> The little egret used to be an exceptionally rare visitor to England. Then, in 1996, a pair bred in East Anglia; there are now around fifty breeding pairs resident in the region. Climate change has been an opportunity for this versatile species to increase its range, drawn across the Channel by the increasingly hot summers of southern England.
>
> (128)

"Ghost Species" thus ends on a small note of hope. The waves of extinction that have engulfed East Anglia have emptied the landscape, and the text is above all a narrative exploration of this loss. But, following the deluge, the white dove has been dispatched and returns with an olive branch in its beak. The world that follows the rainbow may not be the same world as was lost, but there is some hope nonetheless.

"Silt," which appeared in a special issue of *Granta* in 2012 and was subsequently published as a stand-alone miniature publication, also forms a part of Macfarlane's 2012 book *The Old Ways*, which explores ancient pathways and pilgrimage routes, both in Britain and elsewhere. "Silt" is an account of his walking of the Broomway, a tidal causeway off England's east coast.

In "Silt," the narratorial perspective oscillates throughout the text between that of a distanced and informed observer and the subjective voice with its subjective impressions of the landscape. In one single passage a paragraph in length, we join the narrator, who is sitting on the sea wall with his companion, eating sandwiches; we learn that the sun was out; and are invited to share in the physical sensation as the sun dried their bare legs: "The clay dried fast on my legs, crisping the hairs and tightening my skin so that I felt kiln-fired—a mud-man." Then, almost in the same breath, the narrator observes avocets take flight close by, remembers the curlews he saw earlier in the year, and begins to reflect on the migratory routes of birds:

> Staggering recent research into avian navigation has revealed that, by means of retinal proteins called cryptochromes, birds can actually *see* magnetic fields. Magnetic-force structures are visible as darker or lighter forms, which are superimposed on the conventionally visible landscape, and so help guide birds to their destinations.
>
> (Macfarlane 2012, 57)

This shift, from the deeply subjective sensation of drying mud to the almost incomprehensibly sensitive vision of migratory birds, is giddying, and reminiscent once again of Sebald's interest in perspectives, in particular his insistence on the unreliability of any single viewpoint on the world or its history. The mosaic of ways that the landscape is experienced, from knowledge about the way that birds move to the sensation of mud drying on skin, expresses the complexity of the human relationship with the natural world. The system by which birds navigate also introduces the theme of (human) disorientation that is also a component part of the narrative of the Broomway. The birds can find their way via magnetic fields, and this serves as counterpoint to the stories of human walkers. Their difficulties in negotiating routes and estimating distances in the watery landscape, and in avoiding the dangers inherent in making the crossing, permeate the text:

> It was a shallow tidal channel and the water it held caught and pooled in the sun, such that its route existed principally as flux; a phenomenon of light and of currents ... so we walked it northwards, along that glowing track made neither of water nor of land.
>
> (43)

Later in the text, Macfarlane's acquaintance warns him, "It's a weird world out there on the flats.... Nothing looks the same as normal. Gulls can seem as big as eagles. Scale and distance change. It's very easy to lose your bearings" (48).

The narrator explains that the Broomway got its name from the stakes (brooms) driven into the sand at regular intervals to prevent the walker from straying from the path into the mud or quicksand, or losing their sense of direction and being swept away by the incoming tide. While the landscape descriptions are shot through with a sense of disorientation as Macfarlane and his companion walk the "mirror-world" of the causeway, the reader navigates the text by a series of historical and factual references that stake out the course of the text and provide context for the landscape of shimmering mudflats. One such intertextual "broom" is the mention of Dunwich. Macfarlane writes:

> At Dunwich, an entire town was swallowed by the sea over several centuries. Nothing of it now is left, though late-nineteenth-century photos exist of its last towers standing crooked on the beach. Historical data about Dunwich is sufficiently profuse that maps have been made of the former outline of streets, buildings and churches, and their positions relative to the current shore. In this way, swimming off the shingle beach, you can float over invisible streets and buildings: the further out you go, the further back in history you've reached.
>
> (53)

Macfarlane makes clear the way that Dunwich focuses as a nexus of cultural memory, even though it has disappeared entirely from the land, rendered inaccessible by the seawater that covers it. The palimpsest that is "invisible" in the waterscape is made visual here in text, again indicating the way in which text and landscape interact over time. The disappearance of the medieval city was a motif in Sebald's *Rings of Saturn*, and unnamed flooded settlements featured in both Meg Rosoff's novel (the subject of the next chapter) and Jim

Shepard's short story that was explored in Chapter 2; the vanitas motif of human glory vanished under the waves links together textual experiences of the landscape. Other "stakes" to guide the reader along the Broomway move back and forth in time, from the first record of the causeway in 1419, to the 1953 flood, to grisly nineteenth-century tales of travellers swept away by the incoming tide. A longer story is recounted when Macfarlane and his companion step "out onto the flats, and back into the Mesolithic" (51).

Macfarlane tells the story of Doggerland, once the dry expanse of land that linked the eastern edge of England with what is now Germany in the west and the Netherlands to the south. Doggerland was discovered by chance, following an archaeological find—a Mesolithic harpoon—fished up by a trawler in 1931. Further archaeological work made it possible to reimagine the drowned landscape as it would have been:

> Doggerland, then exposed, would have been harsh tundra. But as global temperatures rose, melting ice sent freshwater rivers spinning through the tundra, irrigating and fertilizing it, such that it developed into a habitable, even hospitable, terrain. We know that there were trout in the rivers of Doggerland, wild boar and deer in its oak and ash woods, and that stinging nettles grew amongst its grasses.
>
> (52)

Whilst telling us how archaeologists are working to back-map Doggerland rather in the way that historians have back-mapped Dunwich, Macfarlane evokes a clement landscape that sustained our Mesolithic ancestors. He also tells us of Doggerland's gradual inundation by rising seawaters, slowly driving the hunter-gatherers away to the higher ground on either side. And here, the link he sees between the landscape memory of our Mesolithic past and the ecological changes in the future becomes explicit:

> Considering Doggerland now, it is hard not to think forwards as well as backwards. To those living on the vulnerable east coast of England, drowned Doggerland offers a glimpse of the future. Around the coasts of Norfolk and Suffolk, the land is being bitten back by the sea.... Consulting historical maps of East Anglia, you realize that substantial areas of the region have already joined Doggerland: coastlines have become ghost-lines. In places such as these the undertow of the past is strong—liable to take your legs from you and pull you down without warning.
>
> (53)

Macfarlane works to counter the disorientation of his readers in a landscape by providing landmarks that allow us to move forwards and backwards in time. Imagining the history and the geological past of the landscape is what allows us collectively to make sense of the landscape, to "read" it, cross it safely, and to understand our relationship with it.

The subtext of human (dis)orientation in a landscape is underlined further by Macfarlane's references to maps: cartography, the human attempt to understand and record space, appears throughout the text as descriptions of maps, interjections of a visual tool into the play of words. The introduction of the landscape is by means of a word-map: the second paragraph of "Silt" opens with, "If you consult a large-scale map of the Essex coastline between the River Crouch and the River Thames…" (48). Another map is the hand-drawn one by Macfarlane's acquaintance Patrick Arnold: the back-maps of Dunwich and Doggerland are imaginative maps of places, providing the seascape with historical layers of mutation and meaning. Maps are another feature of the hybrid genre of nature writing, and another way of gaining perspective on place.

It is this layered understanding of a landscape's past—the reading of a damaged manuscript and manifold palimpsests—that allows us to engage imaginatively with our present and future. This kind of imagination needs landscapes; moving between the human past and the human future limits us to an anthropocentric worldview that cannot meet the demands imposed on it by the threat of climate change and environmental crisis. Macfarlane can make the jump to imagining the Doggerland of the future, whether or not this Doggerland will still have any humans to walk its valleys and go fishing in its rivers:

> It is likely that, thousands of years in the future, when the temperature cycles have turned again and the world's water is once more locked up in ice, Doggerland will be re-exposed; filled this time with the wreckage of an Anthropocene culture—a vast junkyard of beached derricks and stranded sea-forts, botched pipes and wiring, the concrete caltrops of anti-tank defences, fleets of grounded and upended boats, and the spoilheaps of former houses.
>
> (54)

We are drawn to imagine the Anthropocene, not by what it means for us now, but by what it will leave behind. In his "Four Theses," Chakrabarty writes:

> To go along with Weisman's experiment, we have to insert ourselves into a future "without us" in order to be able to visualize it. Thus, our usual historical practices for visualizing times, past and future, times inaccessible to us personally—the exercise of historical understanding—are thrown into a deep contradiction and confusion.
>
> (2009, 197)

Inserting ourselves into a future world "without us," as Chakrabarty puts it, is the implicit programme of Macfarlane's writing in "Silt"; his prose is both a Sebaldian lament for all that is lost, and a note of belief in the power of the planet to regenerate over deep time, if not necessarily within the smaller-scale future of its human inhabitants.

Chakrabarty's "futures that we cannot visualize"

In Chakrabarty's second thesis, he writes that "it is not surprising then that the crisis of climate change should produce anxieties around futures that we cannot visualize" (Chakrabarty 2009, 211). The genre of new nature writing is a response to this, to the need to be able to see and imagine possible futures in order to productively engage our anxieties and shift politics in a direction that might mitigate crisis. The ethical position that underpins Macfarlane's work is apparent, not just in the texts themselves, but also in his reflections on them. In 2003, slightly before the revival of the genre that subsequently became known as the new nature writing, Macfarlane answered a lament by Richard Mabey for the decline of writing about nature:

> Why, though, should any of this matter? Literary genres come and go: why should this one in particular be grieved for or celebrated? The answer is that the disappearance of this sort of writing will only widen what Mabey calls the "ominous and growing fault-line in the way we perceive and talk about nature"—and its reappearance might go some way towards closing this fault-line. The natural world becomes far more easily disposable if it is not imaginatively known, and a failure to include it in a literary regard can slide easily into a failure to include it in a moral regard.
>
> (Macfarlane 2003)

A moral regard for nature is a critical tool, according to both Macfarlane and Mabey, in caring about its future. In *Nature Cure*, Mabey notes despairingly that "we have evolved as talkers and dreamers. This is our niche in the world, something we can't undo. But can't we see those very skills as our way back, rather than the cause of our exile?" (Mabey 2005, 37). Mabey expresses the importance of the human imagination in Chakrabarty's argument, in his second thesis, that, "in the era of the Anthropocene, we need the Enlightenment (that is, reason) even more than in the past" (Chakrabarty 2009, 211). The Enlightenment was driven by the aims of reason and progress and this has led to the Anthropocene, but the same drive for reason and progress is, in Chakrabarty's view, also still the only way of mitigating the crisis of this new age. Chakrabarty goes on to express reservations about the way that reason—an understanding of the ecological and geological calamity we face—is translated into politics, stating that "politics has never been based on reason alone": Mabey calls for literary "talking and dreaming" as a counterweight to reason, an emotional responsiveness that might, perhaps, when rendered politically, be part of a new ecological Enlightenment.

Whilst both writers—and implicitly the others who are counted as adherents of the genre, whatever they may think of the nomenclature (Hoare 2015, 56)—evidently believe in the transformative potential of the imagination, Macfarlane is wary of writing that is overly zealous in aiming to achieve a particular effect:

> There are good reasons not to exaggerate the possible consequences of nature writing. One is that it often preaches to the converted: the people it reaches tend to be those with the most developed environmental consciences.... Another reason is that nature writing can often feel too pious and gentle in its urgings—the green equivalent of attending Sunday school. And yet a law of unintended consequences has always governed literature, as *The Monkey Wrench Gang* made clear. Writing that aims to provoke specific behaviours has a name: propaganda. And writing that seeks to provoke specific emotions also has a name: kitsch. Literature, by contrast, does not deal in deliverables. It stirs the sediments of thought and morals, setting them strangely aswirl.
>
> (Macfarlane 2013b, 167)

Macfarlane's assertion here that literature is governed by a law of unintended consequences positions nature writing firmly within the tradition of imaginative writing, despite its use of scientific intertexts. Nature writing makes no claim to didactic strategies (for, by doing so, it would become propaganda or kitsch). Its purpose reverts to storytelling as entertainment (or for aesthetic consumerism, to borrow Tim Clark's phrase)—responsible storytelling, admittedly, of the kind advocated by Thom van Dooren, in which reader and writer are exhorted to ask, "How are 'we' called into responsibility *here and now*, and how will we take up that call?" (van Dooren 2014, 147). The new nature writers are vocal about their political engagement in environmental questions and, thanks to the marketability of the genre, they are raising explicit questions about the roles of humans in the world in mainstream literary culture. Macfarlane again:

> I think this kind of writing is becoming interested in future wilds and future natures: places and times where the wild or the natural is creeping back. We live now in the Anthropocene, which is the era—the name that we've now given to the so-called geological era—in which human influence is the dominant, earth-shaping force. And I think, in a sense, that this is what this issue is about. It is a reaction to living in the Anthropocene. And the pastoral won't do, the idyllic won't do any longer, we can't kind of settle back into the comfy stylistic grass. And instead, this sort of writing needs to be admitting that there's poison in the system—and admitting poison to its own system.
>
> (Macfarlane 2008b, 06:50 mins)

The need for both reason and dreaming, scientific understanding and political will, have shaped a genre that descends from the romantic and the pastoral to mix scientific and emotional responses to the natural world—the nature in nature writing is no longer a retreat from the canker of industry and urban living, but a place to question what kind of habitation humans are capable of, and to make us as humans sensitive to what we have lost in the past and stand to lose in the future.

The landscapes of East Anglia, generated by humans and now threatened by anthropogenic climate change and widespread species loss, prove to be an ideal theatre in which the conflict between the human and the more-than-human ways of living in the world can be played out. The traces of both sudden and slow changes imprinted on the physical landscape and the traces left by literary landscapes in our mind interact to create an East Anglia of the imagination: the instability of the landscape, the history of human engineering, and its vulnerability to the manifestations of climate change are mirrored in writing that juxtaposes a range of different perspectives and modes. New nature writing is responding to a threatened landscape by attempting to find an authentic expression for how we see and interpret the world around us. Its use of palimpsest expresses a complex and multifaceted relationship between humans and their environment and projects our imagination of the past onto an unknown future.

Notes

1 David Gessner spoke of nature writing breaking out of "the comfortable straitjacket of genre." Cited in podcast interview by Robert MacFarlane (2014, 07:30 mins).
2 See the website http://wainwrightprize.com/ (last accessed 27 June 2018).
3 Macfarlane's use of the term "ecology" here calls to mind Hubert Zapf's (2002) definition of literature as cultural ecology, in which literary texts are themselves part of a cultural ecosystem, providing imaginative counterdiscourse, reintegrative interdiscourse, and a cultural-critical metadiscourse. Zapf's definition would be a means of explaining the politicized genesis of the new nature writing and its popularity in a time of environmental crisis. Cf. on cultural ecology also Goodbody (2014, 61–74, esp. 70) and Hennig (2014, esp. 40–50).
4 *Birds Britannica* was named British Birds/BTO Bird Book of the Year in 2005.

References

Barkham, Patrick. 2014. "Should Coastal Britain Surrender to the Tides?" *Guardian*, 7 February 2014. Accessed 27 June 2018. www.theguardian.com/environment/2014/feb/07/should-coastal-britain-surrender-to-tide.

Buell, Lawrence. 1995. *The Environmental Imagination: Thoreau, Nature Writing, and the Formation of American Culture*. Cambridge, MA: Harvard University Press.

Canton, James. 2013. *Out of Essex: Re-Imagining a Literary Landscape*. Oxford: Signal.

Chakrabarty, Dipesh. 2009. "The Climate of History: Four Theses." *Critical Enquiry* 35, no. 2 (Winter): 197–222.

Clark, Timothy. 2011. *The Cambridge Introduction to Literature and the Environment*. Cambridge: Cambridge University Press.

Cocker, Mark. 2005. "Introduction." In *Birds Britannica*, edited by Mark Cocker and Richard Mabey, ix–x. London: Chatto & Windus.

Cocker, Mark. 2014. *Claxton: Field Notes from a Small Planet*. London: Jonathan Cape.

Cowley, Jason. 2008. "Editor's Letter: The New Nature Writing." In "The New Nature Writing." Special issue, *Granta* 102 (Summer): 7–12.

Goodbody, Axel. 2014. "Ecocritical Theory: Romantic Roots and Critical Impulses from Twentieth-Century European Thinkers." In *The Cambridge Companion to Literature*

and the Environment, edited by Louise Westling, 61–74. Cambridge: Cambridge University Press.
Hennig, Reinhard. 2014. *Umwelt-engagierte Literatur aus Island und Norwegen: Ein interdisziplinärer Beitrag zu den environmental humanities*. Frankfurt am Main: Peter Lang.
Hoare, Philip. 2015. "The Naming of the Shrew." *New Statesman*, 20–26 March 2015. 56–8.
Hoskins, W.G. 1955. *The Making of the English Landscape*. London: Hodder and Stoughton.
Iovino, Serenella and Serpil Oppermann. 2014. "Introduction: Stories Come to Matter." In *Material Ecocriticism*, edited by Serenella Iovino and Serpil Oppermann, 1–17. Bloomington: Indiana University Press.
Lekan, Thomas M. 2014. "Fractual Earth: Visualizing the Global Environment in the Anthropocene." *Environmental Humanities* 5 (1): 171–201.
Mabey, Richard. 2005. *Nature Cure*. London: Chatto & Windus.
Macdonald, Helen. 2014. *H is for Hawk*. London: Jonathan Cape.
Macfarlane, Robert. 2003. "The Call of the Wild." *Guardian*, 6 December 2003. Accessed 24 March 2016. www.theguardian.com/books/2003/dec/06/featuresreviews.guardianreview34.
Macfarlane, Robert. 2007. *The Wild Places*. London: Granta.
Macfarlane, Robert. 2008a. "Ghost Species." *Granta* 102 (July): 109–28.
MacFarlane, Robert. 2008b. "Ghost Species: Video." Video interview for *Granta Magazine*, 18 July 2008. Accessed 24 March 2016. http://granta.com/video-ghost-species/.
Macfarlane, Robert. 2012. "Silt." *Granta* 119 (May): 41–60.
Macfarlane, Robert. 2013a. *The Old Ways: A Journey on Foot*. London: Penguin.
Macfarlane, Robert. 2013b. "New Words on the Wild." *Nature* 498 (June): 166–8.
Macfarlane, Robert and Rebecca Solnit. 2014. "Live Event: Robert Macfarlane and Rebecca Solnit on Nature Writing." In podcast interview for *Orion Magazine*, 3 January 2014. Accessed 24 March 2016. https://orionmagazine.org/article/live-event-robert-macfarlane-and-rebecca-solnit-on-nature-writing/.
McVeigh, Tracy. 2015. "Make Way for the Lynx and the Bear as 'Rewilding' Projects Gather Pace across Britain." *The Observer*, 3 May 2015. www.theguardian.com/environment/2015/may/03/lynx-rewilding-britain-wildlife-countryside. Last modified 14 February 2018.
Pretty, Jules. 2011. *This Luminous Coast*. UK: Full Circle.
van Dooren, Thom. 2014. *Flight Ways: Life and Loss at the Edge of Extinction*. New York: University of Columbia Press.
Wakild, Emily. 2014. "The Challenge of Scale in Environmental History: A Small Meditation on a Large Matter." In *Crossing Mountains: The Challenges of Doing Environmental History*, edited by Marcus Hall and Patrick Kupper. *RCC Perspectives* 4, 19–29.
Zapf, Hubert. 2002. *Literatur als kulturelle Ökologie: zur kulturellen Funktion imaginativer Texte an Beispielen des amerikanischen Romans*. Tübingen: Niemeyer.

5 Causeways to the past

Anthropocene and memory in contemporary novels

In this chapter, I will explore two contemporary novels set on the coasts of the North Sea. A number of writers in recent years have set works of fiction in the North Sea lowlands; among those that have received critical attention are Julie Myerson, *Something Might Happen* (published 2003), Margriet de Moor, *De verdronkene* (*Sturmflut*) (published 2005), Jim Shepard, "The Netherlands Lives With Water" (published 2010), Jon McGregor, *This Isn't the Sort of Thing that Happens to Someone Like You* (2012), Uwe Timm, *Vogelweide* (2013), Christiane Neudecker, *Sommernovelle* (2015), and Eva Moraal, *Overstroomd* (2015). The number of titles appearing has been prominent enough to suggest that it is something of a trend (Morrison 2015). Besides these listed titles, there has been a great deal of genre fiction, in particular crime fiction, that makes use of the North Sea landscapes. Writers such as Ann Cleeves, Klaus-Peter Wolf, Haakon Nesser, Elke Bergsma, and others have produced series of novels that are known for their particular regional setting (in Germany, titles are often marketed as a "Wattenkrimi" or "Nordseekrimi").

This contemporary engagement with the North Sea lowlands points towards an engagement with climate change. Scholars both inside and outside the field of ecocriticism have long asserted that literary texts have been slow to engage with the spectre of climate change (e.g., Ghosh 2016; see the Introduction to this book). Chris Maughan, when reviewing Timothy Clark's *Cambridge Introduction to Environment and Literature*, noted that:

> What might also strike readers is the lack of reference to conventional fiction. There is a lot of literary theory, a lot of discussion of environmentalism in a direct sense, but proportionately less straightforward literature connected to it.... It is also perhaps an intimation that any lack is in fact a result of the unprecedented scale of contemporary environmental problems; that is, we are still waiting for literary responses commensurate with the scale of environmental realities.
>
> (Maughan 2011)

However, in *Anthropocene Fictions: The Novel in a Time of Climate Change*, Adam Trexler addresses this perceived lack of engagement, claiming that climate change is more present in fiction than many scholars have allowed for. His "broad, archival approach to climate fiction" (Trexler 2015, 19) looks at trends and across genres to identify the form and functions of climate change in literary fiction. Drawn from the broad corpus mentioned above, the two novels in this chapter, Meg Rosoff's *What I Was* (2007) and Kjersti Vik's *Mandø* (2009) are exemplary "texts of the Anthropocene" (Trexler, 20). While not explicitly environmentalist texts or novels *about* climate change, both novels use their lowlands setting to move between the past and future of the environment, engaging with the questions of time, human agency, and environmental change that are so central to understandings of the Anthropocene.

Meg Rosoff is an American author living in the UK; Kjersti Vik is a Norwegian writer from the southwest coast of the country. Both of them have chosen to set their fictional works on the seaward edges of the North Sea lowlands—Rosoff's teenage protagonists are on a fictionalized East Anglian coast, and Vik's slightly older ones are on holiday on the Danish tidal island of Mandø. The novels lend themselves well to comparison; although Rosoff's novel is marketed specifically as a work for teenagers (Rosoff's reputation in this genre is well established following her 2004 novel *How I Live Now*, which won a number of awards including the British Guardian Children's Fiction Prize and the American Printz Award for young-adult literature) and Kjersti Vik's debut novel is targeted at an adult readership, both works involve young adult protagonists who are establishing relationships with others on the margins of a coastal landscape. Rosoff's narrator is a boarder in a repressive boys' school, St. Oswald's, in the early 1960s; the plot of *What I Was* is propelled forwards when the unhappy schoolboy meets and befriends Finn, another boy who lives alone on an isolated stretch of coast and whose life is free of the boredom, petty rules, and bullying of school. In Vik's novel, which is set in contemporary Denmark and narrated by different characters in turn, the framing narrator is Maja, who has come to Mandø for a week's holiday with six student friends. She encounters Claus, a native of Mandø, and the novel details their brief liaison and the impact it has on both parties and on the rest of their group.

In both novels, the relationships in question cross the divide between what Michael Pearson has called "littoral society" and inland (mainstream) culture. In his 2006 essay "Littoral Society: The Concept and the Problems," Pearson describes littoral societies as those "that have more in common with other littoral societies than they do with their inland neighbours" (Pearson 2006, 353): in essence, communities that are defined by their proximity to the sea and marginality compared to the mainland. This marginal status applies to both the island of Mandø and the fictional island of The Stele in Rosoff's East Anglia. Pearson describes frequent characteristics of a littoral society as being "a symbiosis between land and sea," with indeterminate and often contested borders (354). Littoral spaces are liminal; in the literary works examined here they provide a space for alternative ways of life and worldviews that are missing in inland society.

Pearson reviews various examples of littoral societies from around the world, and ends by noting the transformation to which these littoral societies have been and are being subjected:

> The complex symbiosis between land and sea that we found to characterize littoral society for most of history is fast being transformed. Land influences, often from far away, are profoundly modifying what used to be an important, albeit elusive, segment of human society … It is not a matter of the end of littoral society, but rather that it has undergone, over the last century or so, more major changes, huge stresses, even transformations than was ever the case before this.
>
> (272–3)

This is important in that it frames the marginality of these societies in temporal as well as spatial terms. Greg Dening also writes of beaches as "places for special historical insights" (Dening 2004, 13), and this sense of the link between past and present is something that is relevant in both of the novels discussed in this chapter, and which will be explored further.

The marginal world of the littoral is connected to the normative values of mainland culture in both novels by a tidal causeway. Causeways are landscape features that exist around the North Sea basin, connecting islands and peninsulas with the mainland at certain times of the day. The dangers connected with them are well documented—orientation on an empty expanse of sand or mudflat is often deceptively difficult, sea mists can descend and render visibility poor, and misjudging the timing of a crossing can have fatal consequences—and indeed, they function as a plot device in both of the novels by Vik and Rosoff. The threat of being cut off and the danger of mistiming or disorientation play a part in driving the plot forward.

Causeways have thus proved to be a popular device in literary fiction, and there are many legends attached to causeways: famously, for example, the English King John is rumoured to have lost much of his baggage train to the incoming tide, including the crown jewels, on ordering it to cross a causeway across the tidal estuary of the Wash in the year 1216. Robert MacFarlane's account of walking the tidal causeway called the Broomway in "Silt" was explored in Chapter 4 of this book. Arthur Ransome's wholesome adventurers almost get caught out crossing a causeway in *Secret Water* when they misjudge the time, and they have to be rescued by boat. In "Die Flut ist pünktlich" ("The Tide is on Time") by Siegfried Lenz, a man is led to his death when his watch is altered, so that he mistimes his walk across the mudflats and is swept away by the tide.

The causeway also functions symbolically as a permeable barrier between the margin and the mainland. Passage between these two places is mediated by an understanding of the landscape; it requires knowledge of time and tide, and a respect for the prevailing weather and conditions. The littoral inhabitants know their environment and are able to keep themselves safe, but the mainlanders do not, and this source of tension between the two underlines the difference in human relationships to the landscape. A character dies in Rosoff's novel; in Vik's, a character nearly does.

The relationship between different parts of the landscape—inland, causeway, littoral—is thus an important feature of both works. It marks the transition of the young protagonists into new social and cultural spaces; these geographical margins can be read as heterotopias in the Foucauldian sense, spaces where social norms are suspended or are perceived to be lacking. Jürgen Hasse argues for the *Weltnaturerbe Wattenmeer* as a heterotopia, a mythical space where nature holds sway, a "säkularer Fluchtpunkt" or secular refuge from the modern world, even if this refuge is illusory (Hasse 2011, esp. 102–5). Hasse, writing of the UNESCO-protected nature sanctuary of the Wadden Sea coast, proposes that the heterotopias create a space in which the power and influence of myth can be exerted (105). In both the fictional texts in this chapter, the littoral is, in Foucault's terms, a crisis heterotopia, a place where the adolescent characters can enact their crises outside of the bounds and rules of mainstream society (Foucault 1984, 46–9). However, the crossing and re-crossing of the causeway is also a marker of a temporality that is particularly significant on the fast-eroding coasts.

In both of the texts examined here, the past of the landscape plays a central role. Characters remember the historic flood events and the shape of the coastline and the human settlements that have disappeared beneath the sea: they are trying to interrogate and understand the landscape's history, and the broader implications this landscape has for our cultural moment. Thus the causeway, which links the modern with a littoral that often seems to be a relic of a past way of life, is also a symbolic (and precarious) link between different points in time.

Reflecting on his influential concept of *lieux de mémoire* in 2002, Pierre Nora locates the upsurge of interest in collective and personal memory at the moment when the "acceleration of history" (in the words of another French historian, Daniel Halévy) becomes pronounced. The acceleration of history refers to the spiralling constant change that characterizes the modern world; the absence of stability and the inability to imagine the future call into being a new dependence on the cult of memory. Nora describes it thus:

> An increasingly rapid change, an accelerated precipitation of all things into an ever more swiftly retreating past.... It is of crucial importance, for it has shattered the unity of historical time, that fine, straightforward linearity which traditionally bound the present and the future to the past.
>
> (Nora 2002)

The multiple volumes of Nora's *Les lieux de mémoire* are concerned with the intellectual and national history of the French Republic in its metamorphosis from a country largely dependent on rural agriculture to a modern urban state, and much of the work done in memory studies in recent decades has focused on the sites and memorial processes arising from the major traumas of the twentieth century. Simon Schama links the renewed interest in memory with specific physical environments in his *Landscape and Memory* from 1995, and the concept of disaster memory has also come increasingly to the fore, as we have seen in Chapter 1. The environmentalist movement that started in the 1970s has contributed to an awareness of the fragility of landscapes, and interest in physical place and memory is increasingly being taken up by environmentalists and environmental historians (see for example the special issue of *Global Environment* entitled "Environment and Memory" from 2013) (Uekötter 2013).

The acceleration of history correlates chronologically with the "Great Acceleration" identified by geologists and scientists studying the effects of anthropogenic change on the planet's ecosystem, and thus is also implicated in the idea of the Anthropocene. The *ecological* instability brought about by the exponential increase in the burning of fossil fuels and industrialized agriculture, it can be posited, also created a new cult of *ecological* memory. Time and memory are important aspects of the Anthropocene concept. The science journalist Christian Schwägerl writes:

> The Anthropocene extends the traditional "short now" (ranging from single moments to the duration of an individual life) to a "long now" that includes the effects of our daily lives of today on the centuries, millennia, and actual millions of years to come.
>
> (2013, 32)

Dipesh Chakrabarty made a similar point in his "Four Theses":

> The discipline of history exists on the assumption that our past, present, and future are connected by a certain continuity of human experience. We normally envisage the future with the help of the same faculty that allows us to picture the past. Weisman's thought experiment [imagining a world without humans] illustrates the historicist paradox that inhabits contemporary moods of anxiety and concern about the finitude of humanity. To go along with Weisman's experiment, we have to insert ourselves into a future "without us" in order to be able to visualize it. Thus, our usual historical practices for visualizing times, past and future, times inaccessible to us personally—the exercise of historical understanding—are thrown into a deep contradiction and confusion.
>
> (2009, 197)

All three writers mentioned above—geologist, historian, and journalist—emphasize the significance of historical understanding in the context of facing the future; Schwägerl and Chakrabarty both focus on the endeavour to see both past and future in less anthropocentric terms than has largely been the case up until now. In the two very contemporary novels examined in this chapter, it is possible to see how the understanding of the Anthropocene as a rupture in the continuity of historical thinking is being both reflected and constituted in literary form: the landscape of the North Sea lowlands functions in the two novels as an environmental site of memory, its past a key resource for grasping current and future moments.

Using the questions of timescale and human agency raised by the concept of the Anthropocene, this chapter shows how works of fiction that explore the changing relationship between humans and the environment in the past and the present become an implicit part of a new genre of writing about the crisis of climate change. The fragility of the coastline is more than just a subtext to the works by Rosoff and Vik; while they are not explicitly environmentalist texts, they weave a sophisticated understanding of the timescales of both human and planetary change into their stories of human rites of passage.

What I Was

> I am a century old, an impossible age
> and my brain has no anchor in
> the present. Instead it drifts, nearly
> always to the same shore.
> Today, as most days, it is 1962.
> The year I discovered love.
> I am sixteen years old.
>
> (Rosoff 2007, non. pag.)

What I Was is narrated, we can calculate, from the year 2046, but the narrator is remembering the events of his youth; most of the events in the book happened in the year 1962. Tellingly, the epigraph mixes the idea of space and time—the year of 1962 is represented as a seashore towards which the narrator's thoughts constantly return. The representation in the epigraph of time as landscape (or seascape), and the fact that the epigraph points to both future and past dates, sets us up for a novel that is orientated both towards our future and towards our past. The narrator has a perspective from which he can pass judgement on the events of the past and on what has come to pass since then.

The narrator (who remains unnamed until an almost glancing reference to a first name, Hilary, at the very end of the novel) has arrived as a new boy at his Spartan and disciplinarian boys' boarding school, St. Oswald's. We learn that he has been asked to leave previous schools; that he is a mediocre scholar, a poor athlete, and has no interest in socializing. He has no ambitions beyond survival. He expects to be unhappy at St. Oswald's, and indeed, his first impressions do nothing to dispel this expectation. Standards are low, the architecture is depressing, and bullying is institutional. The landscape is, initially, part of this forbidding scenario; "by the time we reached St Oswald's, fog had completely smothered the coast. Even this far inland, the mist was impenetrable; our white headlights merely illuminated the fact that we couldn't see" (1). St. Oswald's

School for Boys is a loveless universe of its own, isolated from the rest of the country.

But the landscape of littoral East Anglia soon becomes an escape route for the narrator, a relief from the gloomy environs and unfriendly inhabitants of his school. From the beginning, the landscape is presented over time, imbued with a strong sense of history.

> It all began on the coast of East Anglia, past the indentation where the River Ore ran salt and melted into the sea. There, a bit of land stuck out from the mainland, a small peninsula roughly shaped like a rat's nose. In maps (old maps), this peninsula was labelled "The Stele," after a seventh-century commemorative stone marker, or "stele," found very close to school property in 1825....
>
> For only a few hours a day, when the tide was very low, the little peninsula could be accessed via a damp sand causeway. All around it, salt marsh and reed beds provided homes for nesting waders and waterfowl—oystercatchers, little terns, cormorants, gulls—and had once done the same for Roman, Saxon, and Viking settlers.
>
> (8–9)

As hinted at in the epigraph, the environmental implications of looking both backwards and forwards are stated early on:

> One of the more notable facts about the stretch of coastline I have just described is that it is sinking at great speed. This is the sort of fact about which it has become fashionable to panic in the middle of the twenty-first century, when nearly everyone agrees that our planet is on its last legs, but it has been true of this stretch of land for at least a thousand years.... Back then, I might not have agreed. For one thing, I was less interested in geological catastrophe. For another, my contemporaries and I tended to view the future as a vast blank slate on which to write our own version of human history.
>
> (14)

Here, understanding the (environmental) past is seen as relevant for an understanding of the (environmental) future. The narrator, thinking back to his days as a schoolboy, displays a lack of interest in the past of the landscape, which is mirrored by his projection of *human* history in the future—history, he suggests, is a matter for humans to produce on their own. This is rather analogous to Pierre Nora's understanding of history (and the future). But the narrator is forced to revise his opinions. Playing truant from a cross-country run, he meets Finn, a boy of his own age, who lives alone in a fisherman's hut on The Stele, cut off from the mainland at high tide. Finn lives free of formal education and the norms of society, which rule it impossible for a child to live alone without adult supervision. The awkward friendship between Finn and the narrator allows the latter a secret life outside of school, and awakens a new interest in the landscape that Finn inhabits.

The narrator makes repeat crossings of the causeway into the other world of The Stele.

> The tide chart had actually worked and I crossed the causeway on damp sand by the last pink streaks of evening light. I reckoned I had a couple of clear hours to hang about before the crossing flooded again.... The sky was a uniform gray and bled seamlessly into the sea at an invisible horizon. There was no up, no down, no past, no future. Aside from the far-off ghost of a coal boat chugging its way from Newcastle, I could have placed myself in the seventeenth century, or the seventh.
>
> (39)

The more the narrator crosses the boundary between Finn's world and his own, the greater his understanding of the landscape, whereas Finn already knows it well; not just by inhabiting it, but also through the history he has been taught by his late grandmother. Finn explains:

> [T]he high-water mark is advancing quickly.... Geologically speaking, that is. The high tide only began cutting us off in the last decade.... If you compare the coastline now to maps from a hundred years ago, you can see how much it's changed.
>
> (61)

The history of this mutable landscape is the key to understanding its future. As readers (bearing in mind too that this novel is pitched at teenagers), we are encouraged to follow the narrator on his journey: his growing interest in the past is reflected also in his growing awareness of the fragility of Finn's existence on the shoreline against the rising power of the sea.

> "There was once a city nearby, quite a big city, between AD 600 and 1200. I sometimes find bits of it on the beach." ... I knew about the city; it was part of the legend of St Oswald's, famous mainly for the nineteen churches drowned with the rest of the town in a great medieval storm.
>
> (62–3)

The narrator and Finn look for artefacts from the vanished city, which bears more than a passing resemblance to Dunwich ("famous mainly for the nineteen churches drowned with the rest of the town in a great medieval storm"), and later for the remains of a Roman fortress likewise located out at sea. The *vanitas* symbol of the ruined civilizations serves as a reminder of the transience of human existence. This search for the actual sites of memory is an active attempt to connect with the region's past. Making this connection is of increasing importance to the narrator:

> It sounds like one of those horrible clichés, but this barely visible construction of two-thousand-year-old stone made the hairs rise on the back of my neck … the reality of it was heavy and dark and shaped like a thing from a nightmare, covered in barnacles and seaweed.
>
> (113–14)

Attempts to reimagine Dunwich and the Roman fort are hampered by the sea; but conversely, this is an aid to imagining future coastline changes that will relegate the current landscape, including the hated St. Oswald's, to the seabed.

The novel reaches its dramatic conclusion when Finn becomes dangerously ill, and the narrator has to take ever greater risks in order to nurse him. Finn's physical fragility is mirrored by the increasing threat to his home. High tides and storms are a danger, and the receding piece of land where Finn lives is ever more threatened and precarious. The link between the landscape's past and its likely future is becoming ever more pronounced:

> The tide was high, and the sea surged against the foundation of the hut. I thought of ancient civilizations lying underwater, or stacked below the hut in uneven layers, the long-forgotten lives of fishermen, weavers, farmers. Finn's place in the universe was so fluid, he might have belonged to any one of those civilizations, any century leading up to this one. If I closed my eyes for a moment I could even imagine that he had come here through a blip in the universe, that while I was gone he would return (silently, gracefully) to his daub-and-wattle hut and slide easily back into an earlier, more savage world.
>
> (155)

The narrator is able to arrange rescue for Finn, but has to flee school to do so, and elects to stay on alone in Finn's hut. The approaching storm arrives, flooding the hut and threatening to engulf The Stele completely. At the peak of the storm, Reese, a schoolboy from St. Oswald's, tries to swim across the causeway to reach the narrator, who watches helplessly as he is swept away and drowned. The narrator survives the storm in the hut and visits Finn in hospital, where he discovers that she is in fact female; her existence on the margins of society allowed her to flout more social conventions than the narrator had realized.

The narrator cannot completely make peace with his past. He moves temporarily back into Finn's abandoned hut, but has to move on again; "it wasn't long before I had to face the fact that I was living in the sea.… There was no point hanging around waiting for floods" (203).

The subsequent part of this passage contains one of the most explicit references to the future of humans in the Anthropocene.

> Would the entire human race end up drowned and forgotten, buried under waves of melting ice with no one left to dig us up or wonder at what was or what might have been? Which brings me back to the present, where we

> have begun to drown in a sea of our own making. Our idyll is gone, squandered on a headlong rush towards the future. Our lives in this fragile place have taken their toll.... We were wrong about time, wrong to believe it would move forward in an orderly fashion filled with promise and opportunity and progress forever.
>
> (204)

Here, the connection between the history of humanity and its uncertain future is expressed directly.

The novel closes with the narrator, now back in his present of 2046, "bobbing quietly in a reliable little skiff over the past" (206). Finn's hut, the causeway, and the boarding school have all been lost to the sea, prefiguring what is to come. The transience of human life seen against longer-scale ecological change is given literary articulation using the coastal landscape as a site of memory and palimpsest of environmental change, and pointing to an awareness of planetary fragility and the future for humans.

Mandø

Rosoff's novel remains fundamentally centred on humanity: its version of history is human history, and the future is projected only from a human perspective. The past and the future that Chakrabarty and Schwägerl call upon in their explorations of the concept of the Anthropocene must necessarily, however, be on a much longer timescale, moving beyond the immediate history of flooding and erosion of human settlements to the prehistory of the landscape. If we are to comprehend the implications of climate change, we must also comprehend the idea of a planet without humanity, or at least without humanity as we know it; and here the imagination of the early and pre-human past is a crucial tool in the envisioning of such a future.

The novel *Mandø* by Kjersti Vik is set in the present day, and is prefaced by an epigraph taken from the Old Testament, which in English translates as:

> And the Lord said, I will destroy man whom I have created from the face of the earth; both man, and beast, and the creeping thing, and the fowls of the air; for it repenteth me that I have made them.
>
> (Genesis 6:7, King James Bible)

This reference to the biblical flood is, like the epigraph of Rosoff's novel, a way of looking both backwards and forwards at the same time; the biblical flood is not just a historical (or literary) event but also a reference to the future, to the truly catastrophic scale of climate change events brought about by human mismanagement and short-sightedness, and the threat of coming floods.

Unlike Rosoff's narrator, who is telling the story of his own youth from old age, *Mandø* is narrated by the young protagonists in turn. Maja and Claus, whose interactions provide most of the plot, are the main narrators, with the other characters taking one single narrative segment each. Rosoff's narrator has the benefit of hindsight in his narration; here, by contrast, the protagonists are narrating from the moment, with very little sense of a longer perspective on their lives. Landscape memory plays a less significant role for Vik's young adult protagonists than it does for Rosoff's narrator, but Mandø's past encounters with the storm tide are recounted in the first few pages of the book by the unimpressed Maja:

> Fremdeles finnes noen håndfuller fastboende på øya, hvilket er ganske ubegripelig … Mandø ligger cirka seks kilometer fra land og er så flat og nedtrykt at den ville ligget under vann flere ganger i året hadde det ikke vært for dikene som går rundt hele øya og beskytter den mot havet … Prestegården, der vi skal bo, er fra søttenhundre-og-den-tid, og den gangen, før dikenes tid, var øyeboerne forsvarløse vitner til at havet med jevne mellomrom slukte øya deres, slukte alt de eide, rapte stygt før det trakk seg tilbake.
>
> (Vik 2009, 13)

> (There are still a handful of fulltime inhabitants on the island, which is totally unfathomable … Mandø is about six kilometres away from the mainland and is so flat and downtrodden that it would be underwater several times a year if it wasn't for the dikes that go round the whole of the island and protect it against the sea. The old rectory, where we're supposed to be staying, is from seventeen-hundred-and-something, whatever, and back then, before the dikes were there, the island inhabitants were helpless witnesses when the sea periodically swallowed up their island, swallowed everything they owned, and belched it out horribly before withdrawing again.)[1]

The key point to note with Maja's retelling of this history is her inability to understand it. She repeats again and again that she cannot see why anyone would live there, "Hva var det som fikk dem til å fortsette å bo der ute?" (What was it that made them carry on living out there?), and she concludes that "forholdene rundt øya er rett og slett uforenlige med kommunikasjon, og følgelig med moderne liv" (the situation out there is completely incompatible with communication, and therefore with modern life) (13), thus making clear her view of Mandø as a marginal and archaic space.

Maja and her friends, a group of bored and quarrelsome young adults, spend a week's holiday on the island, negotiating their trips to the mainland by means of the tidal causeway, and wishing that they were somewhere different. They are only interested in their environment in terms of what they can use it for—kitesurfing, a place to sit and drink, a place to flirt with death by driving across the causeway after the tide has started to come in. A contrasting worldview is introduced in the figure of Claus, an inhabitant of Mandø who has a deep respect for the natural history of the island and the mudflats. Claus is a keen birdwatcher, and his tracking of the movements of the migratory birds that come annually to the island puts his sense of his environment in a global context:

> Fortsatt strømmer det på med fugler fra nord. Vadehavet er deres viktigste rasteplass under trekket. Over ti millioner vannfugler flyr forbei hvert år … Rødnebbternene flyr nesten fra pol til pol, hvert år, frem og tilbake. Hvor mange kilometer tilbakelegger de i løpet av et liv? 200 000 kilometer? 250 000? Fuglene så verden lenge før oss.
>
> (27)

> (The stream of birds from the north continues. The Wadden Sea is their most important stop on their migration. More than ten million water birds fly past each year … Arctic terns fly almost from pole to pole, each year, there and back. How many kilometres do they cover in the course of their lives? 200,000 kilometres? 250,000? The birds saw the world long before we did.)

Migratory birds are, to paraphrase the poet Ted Hughes, a sign that the globe is still working: "They've made it again/Which means the globe's still working, the Creation's/Still waking refreshed, our summer's/Still all to come—" (Hughes 2011, 28), a motif that has taken on a dark note today, with swift populations increasingly under threat. Claus's understanding of Mandø situates it in the ecology of global bird migrations. This imagination of Mandø as a key part of a global system contrasts starkly with its position of marginal territory for Denmark, as I have argued elsewhere (see Ritson 2018).

Claus's understanding of the past is similarly on a planetary scale, and is contrasted with Maja's limited understanding and rejection of the landscape's history:

> Jeg forteller henne om månens massetiltrekning på vannet. Jeg forteller henne også at månen flytter seg litt lenger vekk fra jorden for hvert år som går. Da kloden vår fremdeles var ung, lå månen svært nær … Jeg forteller henne at tidevannsforskjellen ute på det åpne havet nå er 36 centimeter, hvilket er ingenting når man tenker på hvor stor jordkloden er, men for 4 milliarder år siden, da var tidevannsbølgen omtrent 400 meter høy! … Jeg ser for meg den unge jordkloden, flytende og full av bevegelse, hinsides all bevegelse vi kan forestille oss i dag…
>
> —Men det var vel før det kom mennesker? sier hun.
>
> (121–2)

> (I tell her about the moon's gravitational pull on the water. I tell her too that the moon moves a little bit further away from the Earth every year that passes. When our planet was still young, the moon was much closer … I tell her that the difference between high and low tide out on the open sea is now 36 centimetres, which is nothing when you think how big the planet

is, but four billion years ago there were tidal waves that were 400 metres high! … I can see the young earth, fluid and full of movement, beyond anything we can imagine today…

"But that was before there were humans, right?" she said.)

Claus possesses the ability to imagine the globe across a much larger timescale, on which humanity is a passing species. He describes a planet full of roiling waters, reminding us of the iconic image of the blue planet as seen from space, an image of fragility that influenced the environmental movement in the 1970s (Parak 2013, 53–9). (This image certainly had currency in Norway, where the country's sense of itself as a pioneer of environmental protection on the global stage was cemented by Gro Harlem Brundtland's authorship of the report *Our Common Future* in 1987; see Hennig 2014, 211–19.)

Maja has little interest in this view of the natural world—her understanding is still entirely human-centred; to her, the tides that govern the causeway are subject to "tidevannets logikk, som er mer eller mindre ubegripelig for menneskearten" (the logic of the tides, which is more or less incomprehensible to human beings) (Vik 2009, 14). She may not be able to grasp this logic, but Claus can—and thus he is able to see far more clearly than Maja what the future holds. Maja still holds on to what are essentially nineteenth-century, anthropocentric ideas about nature: it is a challenge to be overcome, a mere setting for human stories.

The difference between the characters' views of planetary history is reflected also in their understanding of the history of ideas. Superficially, Vik's protagonists seem to be acting out a rather hackneyed trope of city-meets-country, in which urban ideas of progress and human advancement (Maja) are set negatively against the time-honoured wisdom of living in harmony with the natural world (Claus). But the reality is rather more complicated, because the natural world holds little promise of accommodating its human inhabitants, at least as far as Mandø and the lowland coasts are concerned.

In his "Four Theses," Dipesh Chakrabarty makes the point that the idea of the Anthropocene throws the ideals of the Enlightenment into question, showing how the struggle for, and advances in the understanding of, human freedom (the fight against slavery, resistance to totalitarian regimes, decolonization) since 1750 have correlated with a period of fossil-fuel-intensive industrialization and increasing geological agency. The Enlightenment ideals of reason and progress are being undermined by geological realities. But, as Chakrabarty points out, these two sides cannot be so neatly separated, as Enlightenment reason is currently the only chance for humanity to change the course of climate change and limit its impact on human populations. This leaves our understanding of the past in doubt: "Is the geological agency of humans the price we pay for the pursuit of freedom?" (Chakrabarty 2009, 210).

In *Mandø*, Maja and Claus are a post-Holocene enactment of city-meets-country, with neither side able to live up to the new rationality demanded by the Anthropocene. Maja sees herself as an atheist and a rationalist; Claus is spiritual and susceptible to the mysteries of nature—Maja's presence on *Mandø* is transitory, she is an incarnation of human freedom to travel and determine her own path, whereas Claus is stationary, seemingly stuck on the doomed island by his love for the place. Their positions are given illustration in evolutionary terms by two of the species of wildlife found on Mandø.

> Claus sier at der er forskjellige måter å tilpasse seg på. Noen beveger seg og blir en del av selve bevegelsen, som fuglene, de flyr der det til enhver tid er best for dem, enten det er varmen i sør, eller lyset i nord. Andre graver seg ned og unngår forandring … Claus forklarer at det er forskjellige muslinger som har gravd seg ned i sanden.… De kan tilbringe hele livet sitt der, kanskje ti–femten år, på den samme flekken, unberørt av både flom og tørke.
>
> (134)

> (Claus says that there are different ways of adapting. Some creatures move and become part of movement, like the birds, they fly to wherever it's best for them to be at any given time, whether to the warmth in the south or to the light in the north. Others dig themselves in and avoid change … Claus explains that there are different kinds of clams that have dug themselves down into the sand.… They can spend their whole lives there, perhaps ten, fifteen years, on the same spot, untouched by either floods or droughts.)

Like the birds and the clams, Maja and Claus are displaying different kinds of adaptive behaviour, attuned to their different environments—and the introduction of this evolutionary model of course also expands the novel's perspective to include species that have, or will, become extinct, forced out of existence by competition from other forms of life. The references to evolution, and to Charles Darwin, serve to position Mandø as a site of memory in the history of ideas—a place where the story of evolution, of planetary life, and deep time, is brought to mind.

Maja and Claus have a conversation about Darwin, a figure who embodies the European rational ideal:

> Jeg spør om han aldri har hørt om Darwin. Jeg er ganske spydig mot ham. Jeg sier at det sikkert er fordi de fremdeles går rundt og frykter naturen at religionen står så sterkt her ute.
>
> (94)

> (I ask him if he's ever heard of Darwin. I'm pretty sarcastic with him. I say that it's probably because they're still running round in fear of nature that religion is so strong out here.)

It is Maja who introduces the figure of Darwin, provoked by Claus's quasi-religious outpouring about the mysteries of creation. But Maja also represents ignorance, whereas Claus has a far deeper understanding of the intricate interconnectedness of the Darwinian natural world, "om hvordan alt henger sammen, hver minste ting spiller en rolle i denne verden ..." (of how everything is connected, every little thing plays its part in this world ...) (94). The boundary between Enlightenment rationality on the one hand and superstition and mystical belief on the other has become blurred, the messages garbled in their passing down to a generation that has a precarious status in a vulnerable world. Claus's homegrown mysticism is perhaps more attractive than Maja's cynicism, but his world is soon to be washed away.

> Hvis havet stiger, er vi ille ute her på Mandø ... Havet skal ikke komme mye nærmere før det bli ubeboelig her ... Vi her forrykket jordens balanse, selve livets balanse.
>
> (115)

> (If the sea level rises, it'll look bad for us out here on Mandø.... The sea doesn't have to get much closer before it becomes uninhabitable here.... We have upset the Earth's balance, even the balance of life.)

The ever-present threat is that of the great flood that will punish human disregard for the planet by sweeping the landscape away.

Maja's crossing and re-crossing of the causeway and her encounters with Claus bring her briefly closer to an understanding of the littoral environment and its relationship with the mainland. For the reader, it builds a bridge between the vast and unknowable natural history of the planet that Claus represents, and the rather limited version of human history that Maja advocates. The causeway, though, is precarious and only sometimes passable: symbolically, it emphasizes the difficulty of passing from one view to another, the fragility of the landscape mirroring the fragility of attempts to link human and natural histories. And amongst the young inlanders there is scant respect for the capriciousness of the causeway:

> På veien tilbake til Mandø ser vi at tida vår er i ferd med å løpe ut. Om få minutter forsvinner låningsveien under vann. Fy faen. Vi kjører som svin ... Bilen nærmest hopper av sted og fyker utover den smale veien som akkurat er i ferd med å drukne. Vi nøler noen sekunder til, og så sier Freja:—Vi gjør det, for faen!
>
> (58–9)

> (On the way back to Mandø we see that our time is about to run out. In a few minutes the causeway will have disappeared under the water. Shit. We drive like animals ... the car almost jumps in the air and flies across the narrow road that is just starting to flood. We have a few more seconds, and Freja says, "Let's do it, for fuck's sake!")

Claus, on the other hand, illustrates his awareness of the dangers of the sea. He can remember the storm flood of 1999, and he notes that:

> stormflo og ekstremt høyvann kommer hyppigere nå enn før … Jorden har alltid vært dynamisk, og dette området har vært spesielt utsatt for store endringer.
>
> (115)

> (storm floods and extreme tides are happening more often than they used to … the Earth has always been dynamic, and this environment has been especially vulnerable to big changes.)

He tells Maja a local piece of folklore, a "vandrehistorie," about a family who are out on the mudflats when a thick fog descends; they are unable to find their way back and drown when the tide comes back (123).

The climax of the novel sees the young visitors to Mandø pretend that Maja has gone far out on the mudflats alone, and appeal to Claus to help them find her. He disappears into the mist and has not come back the following day, when they have to return to the mainland. They cross the causeway for the last time; "Vi får se å redde oss i land før havet kommer og tar oss" (Let's get onto dry land before the sea comes to take us) (186) are Freja's final words, ironically underlining the environmentalist subtext of the novel. The fragile link between the littoral and the inland worlds is left behind. It is an inconclusive ending; it is not clear whether Maja's experiences on Mandø, which nearly end in tragedy, have taught her or any of her group of friends anything new about their places in the natural world.

Navigating the causeway

Inserting ourselves into a world "without us," as Chakrabarty puts it, is dependent on our being able to imagine a past without us as well. The number of literary texts written about these particular, vulnerable coastal lowlands is symptomatic of the way that these landscapes can be used as a place to engage imaginatively with environmental change. In both the novels by Vik and Rosoff, overt references to the current problem of climate change and its likely impact are few and oblique, and neither novel is explicitly a "climate change novel"; but, as we have seen, the crisis in ecological understanding permeates the works, and the protagonists' differing levels of environmental awareness create tensions that carry plot and dialogue. These two works would meet Adam Trexler's definition for "novels in a time of climate change"; as he writes, "In contemporary literature, melting ice caps, global climate models, rising sea levels, and tipping points have altered the formal possibilities of the novel" (Trexler 2015, 13). The settings of Rosoff's fictionalized East Anglia and Vik's Mandø and the plot and characterization in both works would both be unthinkable without the backstory of the Anthropocene.

Meg Rosoff's use of the historical imagination shows how the landscape's past can move us a step closer towards an imaginative engagement with modern-day environmental change. Her use of a fictionalized Dunwich as a *vanitas* motif serves as an emblem for the transience of human endeavour in the face of geological change. The littoral space of The Stele allows the encounter between Finn and the narrator to happen, but both The Stele and access to it are endangered by coastal erosion; The Stele is littoral in time as well as in space, and ultimately is washed away.

Vik extends our cultural memories further back in time, moving towards an imagination of the geological past and exploring the crisis in relating rational humanity to a broken planet. Mandø functions as a site of memory both within the mutable landscape and within the history of ideas; its niche status for certain kinds of wildlife recall Darwinian narratives of deep time and evolution, and the interplay of spiritual and rationalist arguments show the dilemmas inherent in contemporary worldviews regarding the origins and the future of humanity. Mandø is thus threatened in the short term (by increasing extreme weather events and coastal erosion), but the island also highlights longer-term threats to the human species caused by failure to adapt to the new climate of the Anthropocene.

In both works, failure to navigate the causeway between margin and mainland is not just a danger to the characters in the book. Failure to make the crossing between the fast-eroding landscape of the past and the environmental crisis of modernity is also a symbol for the relationship between the fragile past and its uncertain future. It is all too easy to become disorientated or lost, or to mistime the crossing: but ultimately, navigating the causeway to the past represents our best hope of finding a way into the future.

Note

1 All translations of Kjersti Vik's novel are my own, made with the permission of the author.

References

Chakrabarty, Dipesh. 2009. "The Climate of History: Four Theses." *Critical Enquiry* 35 (2 Winter): 197–222.

de Moor, Margriet. 2007. *De verdronkene*. Amsterdam: Contact.

Dening, Greg. 2004. "Deep Time, Deep Spaces: Civilizing the Sea." In *Sea Changes: Historicizing the Ocean*, edited by Bernhard Klein and Gesa Mackenthun, 13–36. New York: Routledge.

Foucault, Michel. 1984. "Des espaces autres: Hétérotopies." *Architecture, Mouvement, Continuité* 5: 46–9.

Ghosh, Amitav. 2016. *The Great Derangement: Climate Change and the Unthinkable*. Chicago, IL: The University of Chicago Press.

Hasse, Jürgen. 2011. "Zur mythischen Funktion deklarierter Natur-Landschaften: Das Beispiel des 'Weltnaturerbes' Wattenmeer." In *Küstenmentalität und Klimawandel:*

Küstenwandel als kulturelle und soziale Herausforderung, edited by Ludwig Fischer and Karsten Reise, 97–113. Munich: Oekom.

Hennig, Reinhard. 2014. *Umwelt-engagierte Literatur aus Island und Norwegen: Ein interdisziplinärer Beitrag zu den environmental humanities*. Frankfurt am Main: Peter Lang.

Hughes, Ted. 2011. "Swifts." In *The Poetry of Birds*, edited by Simon Armitage and Tim Dee, 128. London: Penguin.

Lenz, Siegfried. 2010. "Die Flut ist pünktlich." In *Die Flut ist pünktlich: Meistererzählungen*, 7–16. Hamburg: Hoffmann & Campe.

Macfarlane, Robert. 2012. "Silt." *Granta* 119: 41–60.

MacGregor, Jon. 2012. *This Isn't the Sort of Thing That Happens to Someone Like You*. London: Bloomsbury.

Maughan, Chris. 2011. "Ecocriticism." *The Oxonian Review* 17 (3 November). Accessed 26 June 2018. www.oxonianreview.org/wp/ecocriticism/.

Moraal, Eva. *Overstroomd*. 2015. Rotterdam: Lemniscaat.

Morrison, Blake. 2015. "'Drowned in a Sea of Salt:' Blake Morrison on the Literature of the East Coast." *Guardian*, 31 January 2015. www.theguardian.com/books/2015/jan/31/blake-morrison-literature-imperilled-east-coast-climate-change. Last modified 22 February 2018.

Myerson, Julie. 2004. *Something Might Happen*. London: Vintage.

Neudecker, Christiane. 2015. *Sommernovelle*. Munich: Luchterhand.

Nora, Pierre. 1984–92. *Les lieux de mémoire*. Paris: Gallimard.

Nora, Pierre. 2002. "Reasons for the Current Upsurge in Memory." *Eurozine*, 19 April 2002. Accessed 26 June 2018. www.eurozine.com/articles/2002-04-19-nora-en.html.

Parak, Gisela. 2013. "Our Only World—An American Vision." In *Eco-Images: Historical Views and Political Strategies*, edited by Gisela Parak. *RCC Perspectives* 1, 53–60.

Pearson, Michael N. 2006. "Littoral Society: The Concept and the Problems." *Journal of World History* 17 (4 December): 353–73.

Ransome, Arthur. 2001. *Secret Water*. London: Red Fox.

Ritson, Katie. Forthcoming 2018. "Nordic Nature on the Edge of the North Sea: Kjersti Vik's *Mandø*." In *Nordic Narratives of Nature and the Environment: Ecocritical Approaches to Northern European Literatures and Cultures*, edited by Reinhard Hennig, Anna-Karin Jonasson, and Peter Degerman. Lanham, MD: Lexington Books.

Rosoff, Meg. 2007. *What I Was*. New York: Viking.

Schama, Simon. 1995. *Landscape and Memory*. New York: Knopf.

Schwägerl, Christian. 2013. "Neurogeology: The Anthropocene's Inspirational Power." In *Anthropocene: Envisioning the Future of the Age of Humans*, edited by Helmuth Trischler. *RCC Perspectives* 3, 29–37.

Shepard, Jim. 2011. "The Netherlands Lives with Water." In *You Think That's Bad: Stories*, 38–66. New York: A.A. Knopf.

Timm, Uwe. 2013. *Vogelweide*. Cologne: Kiepenheuer & Witsch.

Trexler, Adam. 2015. *Anthropocene Fictions: The Novel in a Time of Climate Change*. Charlottesville: University of Virginia Press.

Uekötter, Frank, ed. 2013. "Environment and Memory." Special issue, *Global Environment* 11.

Vik, Kjersti. 2009. *Mandø*. Oslo: Aschehoug.

6 Under the North Sea

"Petrospectral" futures

This book has hitherto dealt with human interactions with the North Sea environment; the literary texts have explored humanity's changing sense of its place in the world and its relationship with other species, but the physical categories of human, animal, and mineral have remained largely discrete. In this final chapter, we will look at two texts that explore the enmeshment of biological life with technology and geology in the North Sea. In these speculative portrayals, the North Sea lowlands recede from view as a land*scape*, and the texts use the coasts instead as a vantage point to explore the geological and material basis for North Sea life and culture now and into the future.

The oil that was formed millions of years ago in the Cretaceous and Jurassic periods, under what is now the North Sea, was discovered in the nineteenth century, but systematic exploration and drilling began in the 1960s, once the North Sea had been parcelled up and allocated between the coastal nations. The presence of oil in this place embeds the North Sea lowlands in the local risk culture of oil extraction, and also in the global risk culture of fossil fuel technology and the acceleration of environmental and cultural transformation across the world. The two literary texts in this chapter can be considered expressions of global petro-fiction in Heather Sullivan's definition, featuring "tandem, joint agencies coalescing into hybrid human-petroleum actions" (Sullivan 2017, 414), but they are also texts that are tethered to the North Sea coasts as a nexus of human-petroleum agencies and place. In the context of this book, they show how petro-writing is another aesthetic iteration of the North Sea coasts as liminal places that draw us into the imagination of the Anthropocene.

The two texts employ different conventions of genre and structure to explore the subject of oil in the North Sea. The short story "Covehithe" by the British speculative fiction author China Miéville, about revenant junked oil rigs that rise from the sea and return to shore to breed, plays on a latent fear that oil extractivism has created monsters beyond human control; the long poem "Solaris korrigert" ("Solaris Corrected") by the Norwegian poet Øyvind Rimbereid uses its hybridized language and imagery to ponder our co-evolution with oil technologies. The two authors are both critically acclaimed within their respective genres, Miéville as a writer of fiction (in particular in the New Weird tradition) and essays, and Rimbereid as a poet, and both texts have received critical attention. Miéville's 2015 short story collection *Three Moments of an Explosion*, from which "Covehithe" was taken, was reviewed in *Guardian* by celebrated science fiction author Ursula Le Guin; and Rimbereid's collection of poetry bearing the same name, *Solaris korrigert*, won the Norwegian Critics Prize for Literature ("kritikerprisen") in the year it was published. Rimbereid won the kritikerprisen for his poetry again in 2013—the only times that the prize has been awarded for poetry in more than two decades (see the notes on reception in Lindberg 2007). In addition, "Solaris korrigert" was adapted into an opera in 2013, and turned into a staged performance by Arne Dahl and Sjur Miljeteig for the Norske teatret in 2015 that garnered highly positive reviews (e.g., Selås 2015; Steinkjer 2015).

East Anglia and the global oil industry

Miéville's short story takes its title from the small town of Covehithe on the East Anglian coast. Like Dunwich, much of Covehithe has already been eroded (one tourist website encourages visitors to "visit Covehithe soon before the rest disappears into the sea!"[1]) and it serves figuratively in this text as a liminal space to frame the story, a fragile vantage point from which the human protagonists can see the phenomenon of the living oil rigs returning to land. The man Dughan and his young daughter hide inside the exclusion zone in order to witness this particular rig emerge from the sea by night. In flashback, we learn how Dughan was part of the military forces when the first living oil rig returned to land in Canada some years previously. Attempts to repel the rigs were initially violent and destructive: but their landfall was unpredictable and so rigs were able to stay on land long enough to be observed drilling into the ground. Subsequently, "riglets" (infant oil rigs the size of a human adult) hatched and went back into the sea, a pattern of reproduction familiar to us from species such as sea turtles. The humans in Miéville's story have reached an uneasy accommodation with the giant rigs and their breeding patterns, but the implications of the revenant-rig population—whether their appearance is a form of revenge on humans, and whether this accommodation with human life can be upheld—are unclear.

This short story turns on moments of recognition: the moment the first rig stamps onto land; the moment when Dughan and his team realize that the apparitions from the sea are not mechanical monsters, but living beings that are coming on land to breed; the hatching of the first infant rigs. Dughan recognizes the oil platforms by names and the disasters that sunk them below the sea. Miéville makes use of the uncanny, apparitions that are simultaneously familiar and strange. The rigs are still recognizably the creations of human engineers, but they have become living beings, able to move autonomously; they are bent and twisted from the processes that led to their being scuttled, but they also show the ability to heal and grow, "bursts from the derrick known to have been destroyed in the rig's last moments, now heat-twisted but regrown, flared above the forest" (Miéville, 345).

The rigs are hybrids, a mixture of biological agency and human technology: "the towerwork was on a platform. In the glow of the thing's own flame they saw edificial flanks, the concrete and rust of them, the iron of the pylon barnacled, shaggy with benthic growth now lank gelatinous bunting" (341). In the descriptions of the rigs, Miéville juxtaposes rust and concrete with seawater and sealife, and blurs the semantic boundaries between the technological and the biological. The rig "hesitated" and "licked" the air with its flame, verbs that make clear its animal agency, while still composed of the trappings of human engineering and described with a genderless pronoun, "pipes dangled from its underside, clots of it fell back into the sea. It wore steel containers, ruins of housing like a bad neighbourhood, old hoists, lift shafts emptying of black water" (341). The traces the oil rigs leave behind, "the rainbow-filmed liquid spoor, the tripod crater prints" (345) are a mixture of animal and machine.

These hybrid beings are reflected semantically in the text in awkward neologisms that seem to represent an attempt in language to pin down and categorize that which resists categorization. "Misbehavicious" (339) is Dughan's daughter's own word for the unstable edge of the land in Covehithe, so close to the lively agency of the sea, and later terms like "petrospectral" (345) and "ovipositor" (351) mesh the natural and the technological. The language is measured, scientific, careful to cast no judgement on phenomena but only to describe, pointing to an attempt to regularize the observations in the language of science, even as language itself has to be redesigned to keep up.

Miéville's protagonist Dughan remembers political discussions about how to deal with the oil rigs, and these discussions were hybrid too in their knowledge bases, involving not just scientists and engineers, but also "theologians and exorcists, soldiers, veterans like Dughan of those first encounters." This unlikely group shares knowledge to try to understand "what kind of economies of sacrifice were being invoked, for what this was a punishment" (344), but they cannot understand more than the discomfiting fact that the scuttled rigs that had disappeared below the sea to rust away out of sight are returning and re-imposing themselves on human civilizations. The late science fiction author and essayist Ursula Le Guin, writing about "Covehithe" in *Guardian*, points to the link between the huge, rusty machine-beings lurching out of the sea and the fact that the oil industry was created and accelerated by humans; "such a playful yet deeply disturbing reference to the ill times we have brought on the world arouse not a pleasurable make-believe shudder, but the real fear we'd rather pretend we don't feel, a fear that is not simply irrational" (Le Guin 2015). Are the rigs monsters of our own making, the malevolent creation of humanity's heedless extractivism? An expression of a thirst for oil that has overstepped an ecological boundary, "reminiscent of cetacean strandings or aberrant avian migration patterns, as symptoms of an ecological system out of kilter, the resurfacing oil rigs are presented as animomorphized embodiments of an exhausted petrocarbon project" as the scholar Moritz Ingwersen (2017) has put it?

"Covehithe" anchors the landscapes of the North Sea—specifically, a crumbling cliff on the East Anglian coast—in the global landscape of this petro-carbon project. The revisitor rigs, we learn, do not necessarily re-emerge from the waters in which they were wrecked, but are evidently globally mobile; the Petrobas P-36 is identified by Dughan as having sunk in 2001 in the Roncador oilfield off the Brazilian coast. Despite this globality, Miéville's story starts and ends with East Anglia, and with a *vanitas* motif now familiar to us in connection with this coast. Covehithe's medieval church on the clifftop was allowed to fall into disrepair in the seventeenth century and was replaced by a smaller building inside the ruins of the original walls. The spectacle of the small church inside the exoskeleton of the older one fascinates Dughan's daughter on her way to the clifftop at the beginning of Miéville's story: "He stopped her by the roofless ruin of the church, pointed, heard her gasp. She stared while moonlight got past the clouds to the holed and broken walls, onto a low newer church inside the nave of the old" (338). At the end of the story, Dughan looks back along Covehithe Beach in the direction of the church where "services continued within the medieval carapace, remains of a grander church fallen apart to time and the civil war and to economics, fallen ultimately with permission" (351). In his use of the word carapace, Miévelle once again uses a biological term for an anthropogenic structure, and invokes the superstructure of the animate oil rigs. They are equated to the church as a project of human ambition subsequently fallen prey to changing times. This final image, of the hollowed-out church on the crumbling cliff in Covehithe, suggests that the remains of global petro-culture will leave a mark on the landscape that will outlast its significance for human culture.

Solaris corrected

Ingwersen notes in his analysis of "Covehithe" that the oil rigs arrive "… from the bottom up: a return from the bottom of the ocean or the crevices of a dusty floor. The unthinkable other comes from the interior, it has always been part of the system" (Ingwersen 2017). This remark can be extended to Øyvind Rimbereid's "Solaris korrigert," with particular reference to the title, "Solaris Corrected." *Solaris* is the name of Polish author Stanisław Lem's 1961 science fiction novel, well known through its film versions, including one by Andrei Tarkovsky. In the book and the films, Solaris is the name of the mysterious sentient planet studied by Earth scientists; in his novel, Lem emphasizes the unknowability of the planet and the arrogance of humans who attempt to understand it. Rimbereid's title "corrects" the extra-planetary focus of this science fiction intertext, pointing to the unknowable in the depths of the Earth rather than outside of it. Narrated from a high-tech future, in which natural landscape as a category has almost disappeared, Rimbereid sites his poem on the shores of the North Sea, on the site of a city (Stavanger) the development of which is deeply entwined with the petroleum industry. Transporting Lem's Solaris back to Earth points the finger at the human arrogance that has driven oil extraction and technology.

The 800 lines of "Solaris korrigert" are written in a hybrid language that is a mixture of standard Norwegian, local dialect, and other languages from around the North Sea. Audunn Lindholm describes how this is a geographically defined combination of languages, pointing out that Rimbereid has drawn a circle around Ekofisk, the first platform complex in the middle of the North Sea, and created a synthesis of Stavanger dialect, lowland Scots, English, Dutch, and Danish, with Old Norse forms mixed in (Lindholm 2008, 356).[2] This combination of languages, dubbed variously as "framtidsspråk,"[3] "samnordsjøisk" and "solarisk" (Refsum 2010; Selås 2015; Pedersen 2015), throws down the gauntlet for the reader; "to read the poem, you have to go up a level of abstraction," as Claus Madsen asserts (2015, 302). The language creates a gap between the reader and the narrator of the poem, and requires an intellectual commitment from the reader to bridge it. The hybrid language evokes a North Sea culture common all around the coasts, but here one defined not by the shared risk of managing water, as we found in earlier texts (e.g., the Hans Christian Andersen novel in Chapter 1), but around oil.

The following short extract displays the main linguistic features of the poem. I have attempted a literal rendering of the poem in standard British English as a guide to readers, but this inevitably cannot capture the strangeness of the original text for its Norwegian audience. The poem is divided into 42 verses of unequal length, divided by slashes. New sentences are signalled by the first word being given in capital letters.

> SIDDY Stavgersand, siddy min,
> exist nearli ne at all, mang
> konklud naw. KAN henda den exist
> onli som ein olda namn,
>   som symbol an ein siddy?
> JA, som om places i 2480 forts sku wiktig vera!
>   (Rimbereid 2005, 12)

> (CITY Stavgersand, my city,
> exists nearly not at all, many
> conclude now. PERHAPS it exists
> only as an old name,
>   as a symbol of a city?
> YES, as if places in 2480 are still of any importance!)

Norwegian words are recognizable, but the vowels *å*, *ø*, and *æ* have been lost from the alphabet, and syllables have been elided or lost (*mange* ["many"] has become *mang*, Stavanger has become Stavger and joined with Sand, a corruption of nearby Sandnes, as the cities themselves have merged). The idiosyncratic punctuation, the contraction of familiar words, and the capitalized initial words add to the distance created by the unfamiliar language, and the effect is reminiscent of reading medieval illuminated texts with their prominent initials (a feature that links "Solaris korrigert" back to the Icelandic sagas that have been formative for Norway's sense of itself). English words and expressions are used throughout to replace Norwegian ones—"exist nearli ne at all," for example is closer to the English than the Norwegian equivalent phrase "existerer nesten ikke i det hele tatt." The Norwegian used is based more closely on the *nynorsk* written form that is used on the west coast of Norway, where the poem is set—the narrator refers to himself as "aig," for example—and many question phrases in the poem start with the word "om," forming a kind of conditional, which is a characteristic specifically of Stavanger dialect (e.g., "OM dei ku expandera,/bli til total univrs?") (32). The word order of the final line in the extract above "JA, som om places i 2480 forts sku wiktig vera!" is reminiscent of German, with the verb moved to the end of the sentence following the use of an auxiliary verb; the spelling of *wiktig* (in Norwegian *viktig*) also suggests German influence.

The evolution of language that Rimbereid evokes in this mixture of influences does not allow us to slip into familiarity with the text; we are constantly put in mind of the time that has passed between the present and 2480, when the poem is set, and this implies a society in which manifold other changes have taken places, as Lindholm also points out:

> Men når diktet beskriver de historiske forandringene som har funnet sted—forandringene som språkføringen hele tiden minner en om—ansporer det til en refleksjon over hva som kan skje når vi gjennom vitenskapen og teknologien endrer de betingelsene vi lever under.
>
> (2008, 357)

> (But when the poem describes the historic changes that have taken place—changes that the poem's use of language are constantly calling to mind—we are provoked to reflect on what can happen when knowledge and technology lead us to change the basic conditions we live in.)

Science fiction (SF: readers with strong preferences about terminology can apply the abbreviation to speculative fiction, or indeed speculative fabulation, as Donna Haraway riffs in *Staying with the Trouble* [2016]) has long been an aesthetic form that has encouraged reflection, not just on the future, but also on the relationship between humans and the natural world, and there is a growing body of ecocritical scholarship devoted to the genre. In his provocative book *The Great Derangement*, Amitav Ghosh rejects science fiction as a genre that is better able to address the pressing "unthinkable" issue of climate change, seeing the problem in the fact that SF is always set in an "imagined 'other' world apart from ours" (Ghosh 2016, 72). But there is a robust counterargument that the other worlds created in SF function precisely because we are able to imagine them in relation to our own world. In a chapter on the relationship between science fiction and ecocritism, Patrick D. Murphy points to the concept of *extrapolation* as key to understanding how science fiction writing is used to provoke readers to think about the present-day world, pointing out that this provides the connection between science fiction and "nature-oriented literature":

> Certainly, SF is not nature writing, in the sense of that genre's definition as being scientifically based, personal observation written in nonfiction prose. What it can be, however, is nature-oriented literature, in the sense of it being an aesthetic text that, on the one hand, directs reader attention toward the natural world and human interaction with other aspects of nature within that world, and, on the other hand, makes specific environmental issues part of the plots and themes of various works. SF also at times shares with both nature writing and other forms of nature-oriented literature detailed attention to the natural world found in the present as well as to the scientific disciplines that facilitate such detailed attention.
>
> (Murphy 2001, 263)

Science fiction uses alternative or future worlds similarly to the way that historical fiction uses past worlds, in order to reflect and provoke thought on the current moment. Extrapolation provides the critical framework that regards the connections between fictional and real worlds: SF, insofar as it creates fictional environments for its fictional protagonists to exist within, can be a comparative vehicle to think about ourselves as humans and our environments.

Rimbereid's poem includes many of the features described by Murphy—it encourages reflection on landscape, on humans and nature, and on advances in science and technology. Like the texts discussed by Murphy in the cited chapter, "Solaris korrigert" is not reliant on an alibi of fantasy, but is a meditation on a changed environment that is the consequence of decisions made in the world we inhabit today.

The poem is set in the year 2480, in a technicized and hierarchical society of humans and robots on the coast near Stavanger. The Solaris of the poem is a new world that is being created in an empty oil well on the bottom of the North Sea. The narrator has the task of overseeing the robots that are sent below the sea to repair the pipes. In the course of the poem, we learn various details about the narrator's existence and his/her internal thoughts. Although it is not made explicit in the poem, it is at least a heteronormative temptation (particularly in the context of science fiction, which has not, on the whole, been especially forward in overcoming binary gender roles) to see the narrator as male; in this chapter, I have tried to avoid using gendered pronouns, but where unavoidable have used "he" and "him" to refer to the narrator. (Elin Lindberg, while also noting that the narrator is not clearly denoted as male in the text, chooses to use the male pronoun throughout her study of the poem [Lindberg 2007, 5].)

Nature and a sense of place in Stavgersand

In the mention of Ekofisk in her article, Lindholm makes clear the centrality of oil to the Norwegian understanding of place as far as the coasts of the North Sea are concerned, and indeed, Per Thomas Andersen traces the story of Norway's relationship with oil through much of Rimbereid's later work, seeing the *Solaris korrigert* as part of an oil triptych with *Herbarium* (2008) and *Jimmen* (2011);

"Rimbereids fortelling handler om Norge in oljealderen" ("Rimbereids tells the story of Norway in the age of oil," Andersen 2018, 10). The Norwegian oil industry started in 1971 with the Ekofisk oilfield, and despite its rapid evolution into Norway's largest industry (replacing fishing) and the prosperity that has followed in its wake, its rise has been controversial: environmental organizations in Norway are opposed to further oilfields due to the risk of pollution and the effects on fish stocks and tourism (Hennig 2014, 217–18). Oil, and the fossil fuel industry it represents, sits uneasily with Norway's concept of itself as an "ökologische und humanitäre Großmacht," a place of unspoilt nature and *friluftsliv* (outdoor pursuits) (Hennig, 214). In 2018, the Norwegian energy giant Statoil changed its name to Equinor, replacing the big oil and big state associations of the former with the socially desirable image of an equitable, equal Norwegian culture.

By 2480, in Rimbereid's vision, the natural world as we know it is depleted, although the exact causes of the damage are not made explicit. There are still animals in the landscape, but they have become fewer:

> SHI … ein vett, glinsen
> dyr
> seer an meg, fra kaien
> reit vid robot-telten min.
> DYREN must up fra seaen haft
> kommat. EIN oter den must vera.
> DEN staar der vid skinnande augr
> i veik, dusti morgnings-ljus.
> SO big den er!
> SELDAN so big dyr i fri ou see.
> FUGLAR og hundar smaa.
> MOSKITO mang naw an hot hot sumar.
> (22)

> (SHI … a wet, glistening
> animal
> looking at me, from the dock
> right next to my robot-tent.
> THE animal must have come from the
> sea. AN otter it must be.
> IT stands there with shining eyes
> in the weak, dusty morning-light.
> SO big it is!
> RARE to see such a big animal in the wild.
> BIRDS and dogs small.
> MOSQUITOES are many now in the hot summers.)

The human inhabitants of Stavgersand live in numbered sections, "organics," and these are organized to be self-contained and discrete from each other.

The narrator mentions the small cats that also still exist, "DEI krysser sirkl an sirkl/i siddyen, ogso gennom organic 3.4 og 3.5,/forbidden kryssar, undr sonar-/systm og over hoy murar" (THEY cross ring from ring/in the city, also through organics 3.4 and 3.5/forbidden to cross, under zone-/system and over high walls") (22). Thus we learn that the remaining wild species live outside of and despite the barriers that human society has erected.

This lack of animal life in the wild notwithstanding, the idea of nature, of wilderness, still seems to hold humans under its spell. The narrator is enthralled by the otter he sees, which provides a recurring motif in the poem, and Stavgersand is a tourist base for those wishing to experience something akin to a "real" landscape:

> HER er so big flow af turists,
> fra Asi og Russi ofts, som gennom her
> og til fjells ska fara, til green-
>   life i WILD-BEAUTI-PARK-NORWG!
> EIN veeka undr himml
> i fjellparks vid fuglar af seldna sort,
> og kan henda ogso ein litl hjort dei seer,
> som plutsl millon trer er,
> vid ein so rolig, vondrande blikk an dei?
>   i sol, vind, sno, men ofts onli regn:
> DET er wat die kommen for, betalen for ou faa stiga inn i.
> BETALEN for at deirs kropp og breyn an ein *wirklig* place
>   haf verat!
>   (12)

> (HERE there is such a big flow of tourists,
> from Asia and Russia mostly, who pass through
> to go to the mountains, to the green-
>   life in WILD-BEAUTY-PARK-NORWAY!
> ONE week under the sky
> in the mountain parks with birds of rare breeds,
> and perhaps also a little deer they might see,
> which suddenly appears between the trees,
> looking with a calm, wondering expression at them?
>   in sun, wind, snow, but often only rain:
> THIS is what they come for, pay for, and want to experience.
> TO PAY for their body and brain to have been in a
>   *real* place!)

Rimbereid reveals some of the considerations of the sense of place that inform his work in his essay "Om den topografiske diktet" ("On the Topographical Poem"), published in 2006. He talks about the influence of his childhood home in Stavanger, and about the prevalence of "ikke-steder," non-places in the sense of

Marc Augé (Augé 1995) such as the globally ubiquitous McDonalds. In Louise Mønster's definition of Augé's term, non-places are "steder, hvor de stedlige kvaliteter er underdrykt … Ikke-steder er steder, der gør det svært for mennesket at forbinde sig meg sin omverden" (places in which the qualities of place are suppressed … non-places are places in which humans have trouble connecting with their environment) (Mønster 2009). Rimbereid, in his essay, writes:

> Men om de tyngste økonomiske, teknologiske og kulturelle kreftene arbeider hardt for å abstrahere det fysiske stedet, og også evner dette ideologisk, ved å hevde at det mest verdifulle stedet er et slags modernistisk ikke-sted, så holdes stedet like fullt fast av andre, kontekstuelle krefter. Naturen og det reint fysiske vil alltid forbli en taus kritikk av tanken om det ikke-stedlige. Bergens sju fjell lar seg ikke utradere i en håndvending.
>
> (2006, 82)

> (But if the strongest economic, technological, and cultural forces are working hard to abstract physical place, and also to flatten it ideologically by attaching the most value to places that are a kind of modernist non-place, place is holding on equally strongly to other, contextual forces. Nature and the purely physical will always be a silent criticism of the idea of the non-place. Bergen's seven hills can't be erased overnight.)

The landscape features in "Solaris korrigert" display precisely this ability to serve as a challenging counter-narrative to abstracted space (e.g., the organics, which have a number rather than a name as if to underline their abstraction); the landscape of Stavanger and Norway's western coast is recognizable even through the bewildering non-places that make up the narrator's world. The "silent criticism" is provided by figurations of the natural world—the repeated mention of seagrass that moves with the sea's waves, the mud, the sleek form of the otter glimpsed in the harbour, the cats that move freely—that break into the impersonal, numbered parts of Stavgersand.

This tension between the dystopian non-place and the historical place and landscape echo that which we find in W.G. Sebald's *The Rings of Saturn* (see Chapter 3 of this book), between the threatening globalization of modernity and the traditional rootedness of local place. Agnew writes that "place is often associated with the world of the past and location/space with the world of the present and future" (2011, 319), which underlines Rimbereid's futuristic sense of a place that has almost been erased, so completely has it been taken over by technology. In "Solaris korrigert," the failure of place—western Norway's rocky coasts, stormy seas, and steep mountains—to disappear completely serves as a critique of the technological system that has brought this new world about. The seagrass, the mud, and the otter remind us specifically of the natural environment that preceded organic 14.6 and the other quarters of Stavgersand, but they do not present us with a "pure" nature completely untouched by human/technological influence. As we have already seen, the wild cats move between the human-occupied spaces of the organics; the narrator's robots, for example, come up from the seabed coated in mud (18). This literal muddying of the divide between the human, the technical, and the natural is part of a post-human understanding of the future of place:

> Geographically, post-humanists question attempts to purify place, space, or territory, as if they can be—or be seen to be—either wholly social or wholly natural. The ethical implication of this is not to reject anthropocentrism in favour of ecocentrism, but to attend more closely to the moral connections between humans and non-humans arising from their ontological imbrication. This muddies the waters supposedly separating environmental ethics and human ethics.
>
> (Castree *et al.* 2013, 141)

Rimbereid's futuristic vision of place and space underlines the centrality of oil to the North Sea region, the extent of anthropogenic damage, what he sees as an increasing takeover of place by abstract space, and the breakdown of "pure" categories of natural verses cultural. This latter aspect especially—the way in which the biological, the technical, and the cultural have merged—is something that is of particular significance in this poetic representation of the future of this landscape.

"Wat vul aig bli?" Humans in Stavgersand

The first line of "Solaris korrigert" is a question to the reader, "Wat vul aig bli/om du ku kreip fra/din vorld til uss?" (What would I be/if you could creep from/your world to ours?) (9). This direct appeal immediately raises the question of who is speaking. Is this really a fellow human, communicating in a language that is comprehensible if unfamiliar? The poem goes on to suggest that the narrator might be ashamed ("skeimfull") if his addressee were to discover what humans have become. We are given to understand that human life has changed fundamentally. The author underlines this by allowing not just the human individual to speak through the poem's narrator, in his strange tongue, but by presenting us views of human society in Stavgersand Siddy (city).

The poem's presentation of a humanity that is simultaneously recognizable and deeply unfamiliar—uncannily so—pushes us to consider what humans really are. In doing so, we brush up against Dipesh Chakrabarty's third thesis, on the challenge of humanity comprehending itself as a species whilst not losing sight of the differences between individual humans and human societies. Chakrabarty wrote that:

> the consequences [of global warming] make sense only if we think of humans as a form of life and look on human history a part of the history of life on this planet. For, ultimately, what the warming of the planet threatens is not the geological planet itself but the very conditions, both

biological and geological, on which the survival of human life as developed in the Holocene period depends.

(2009, 213)

W.G. Sebald's *The Rings of Saturn*, with its entanglements of human history and the histories of non-human species such as herring and silkworm, gave readers an oblique view of humans as a biological species. Here, Øyvind Rimbereid, imagining the future of humanity in the post-oil Anthropocene, encourages a kind of species thinking by framing his fictional human society by analogy with other biological species. Stavgersand society, we learn, is made up of different kinds of humans—those with clear work-related functions, like the narrator, then the workless "drifters," and the elite "skuggar" (shadows) (23), all of whom relate to a kind of brain at the centre that exerts a force like the gravitational pull of the sun.

The way this society is described, with its clear hierarchy and functionalization of the workforce, calls to mind an insect colony, and indeed, ants are mentioned by the narrator, when talking about his robot workers: "DEI arbeiden/i unbegrensa links, miner robots,/so om ein af dei sku i error gaa,/ein odder kommen, erstatt. DEI er lik maur,/onli konstructet" (THEY work/in unlimited connectivity, my robots,/so that if one of them should go wrong,/another comes, replaces it. THEY are like ants,/only constructed) (25). The sections in which the human inhabitants live are called "organics," which might also put us in mind of the word "organism."

Rimbereid thus implies a human society that is a kind of superorganism, linked by a single form of intelligence. This is, first, very much in the tradition of science fiction in its adoption of ideas drawn from science and applying them in a fictionalized setting; but it also suggests a repositioning of the way we think of humanity vis-à-vis the rest of the biosphere. Humans and robots in Stavgersand live in symbiosis; together they form a "system" that functions in a similar way to insect colonies such as those of bees and ants. This creates an ecocentric view of Earth, with humans subjugated to a part in an ecological-technological system; the narration of this state by an idiosyncratic narrator reproduces the tension between humanity-as-species and the individual human identified by Chakrabarty. Human ingenuity in engineering and controlling the environment is likened to that of ants in an anthill; and if we think beyond this analogy, we can see the anthill as a small part of a much bigger system. Just as ants will recover and rebuild when their anthill is kicked over by a heedless boot, humans adapt themselves to forces well beyond their control and comprehension.

Consider this description by Edward O. Wilson, author of *In Search of Nature*, of an ant colony that functions as a superorganism:

> Another reason that ant social life succeeds is that colonies are able to maintain efficient climate control within the nest. I like to refer to an ant colony as a factory within a fortress. The queen and the nurse-workers

> busily raise the young and increase the population by rapid multiplication, while around them is the fortress-like nest, constructed to hold off enemies. They are protected by the often very aggressive work force, which in many species includes a specialized soldier caste. The nest's structure creates a well-regulated greenhouse environment that permits the factory to operate at maximum efficiency. These ants also control large areas around the nest, from which they harvest food. Furthermore, they are able to bequeath the nest, which is very expensive to produce in terms of energy, together with the territory, to later generations.
>
> (1991, 17)

There are many features of Rimbereid's poem that bear similarities to Wilson's observations of this ant colony—the division of living quarters and the labour hierarchy, the presence of a central control element (the ant queen/breyn-machin BK2884), even the "well-regulated greenhouse environment" that is being built in Rimbereid's *seifa botten*, or Solaris. Technology and humanity have merged, but this kind of symbiosis is as much science as fiction, drawn from behaviour seen in insects.

Rimbereid's superorganism as it is taking shape in "Solaris korrigert" has at its heart the "breynmachin BK2884," a giant chemical brain housed in an empty oil well beneath the sea that has to be fed with neurons:

DEPT I Nordsea
i et emti oilbrunn, 1 km undr sea,
fins allrede Breynmachin BK2884, *hidden og protectat*.
BK2884 haf so big chemical breyn,
spred out millom point og point,
millom nevron-nett og nevron-nett
accumulering seg self i seg self.
WI onli must mata den vid nevrons.
BK2884 er den best master
til ou biobalansera biosferen,
og den best master af oren
ekonomical vorld, af taxes, trafficky, siddyplans
og best master af oren plans for future.
OREN organic- og siddy-
konnections kan ne lefa vidout denna breyn, dei seis.
STOPS BK2884 =
BIG risk for oren praktical vorld.

(30)

(DEEP in the North Sea
in an empty oil well, 1 km under the sea,
we already have Brainmachine BK2884, *hidden and protected*.
BK2884 has such a big chemical brain,

spread out from point to point,
from neuron-web to neuron-web
        accumulating itself in itself.
WE only have to feed it with neurons.
BK2884 is the best master
to biobalance the biosphere,
and the best master of our
economic world, of taxes, traffic, cityplans
        and the best master of plans for future.
OUR organic- and city-
connections cannot live without this brain, they say.
STOP BK2884 =
        BIG risk for our practical world.)

BK2884 is presented here as the epicentre of the colony or hive. But another queen bee figure is the mysterious Mrs Chan, who gives orders in the narrator's quarters, "organic 14.6," without ever being seen by the narrator:

MRS. Chan er 123 age, dei seis.
OG difor smart, vid big human novledg.
        SELF om aig iblant tenk hu ne exist.
AIG tenk hu mang er, er ein sort firm.
OR kann henda hu ne human er,
men onli ein konstructet breyn
        som ein human tenk?
(37)

(MRS. Chan is 123 years old, they say.
AND therefore smart, with big human knowledge.
        EVEN if I sometimes think she doesn't exist.
I think that she's many people, a sort of firm.
OR can be that she's not human,
but only a constructed brain
        that thinks like a human?)

Continuing the analogy of the insect colony, the workers in Rimbereid's colony are both human and non-human. There are robots, who, since they don't need to dream ("ROBOTS treng ne draumar" [19]), are the perfect labourers.

        MEN ven miner simpl robots
arbeiden an pipe-
systm undr sea, ofts om natt
undr moon klar: DEI er total fri!
        KAN robots fri vera?
JA, dei kan.

VEN dei repairat ein problem dept i slam haf,
nearli ein impossibl problem,
og aig kontrollen dei back til kai
og aig seer dei, firkanta og forslamma up af seaen
stiga.
DA dei lucki er.
ONLI dei veit det ne...
(18)

(BUT when my simple robots
work on the pipe-
system under sea, often at night
under moon clear: THEY are totally free!
CAN robots be free?
YES, they can.
WHEN they have repaired a problem deep in the mud,
a nearly impossible problem,
and I command them back to the quay
and I see them, angular and muddy, out of the sea
they rise.
THEN they are happy.
ONLY they never know it...)

Robots are contrasted with "drifters," humans who perform menial tasks for little reward. Unlike the robots, the caste of drifters is seen to be problematic.

... MOST drifters.
DEI som lefr unspesific. DEI haf ne seifa
systm umkring seg.
DRIFTERS haf ne ansikt,
aig somtiims tenk. DEI haf onli
kroppar som dei profar
ou gera meir og meir specific.
(13–14)

(... MOSTLY drifters.
THEY who live unspecifically. THEY have no safe
system protecting them.
DRIFTERS have no faces,
I sometimes think. THEY have only
bodies that they try
to make more and more specific.)

The need to be *specific* is critical and seems to be the condition for being able to move to Solaris, which is described as "den nyast og best mirror-/vorld, den nyast og most healthy af all!" ("the newest and best mirror-/world, the newest and most healthy of all!") (35). It is not clear whether Shiri, the narrator's girlfriend, will join him in Solaris: she is a photographer, an artist, who produces pictures that are "nearli som draumar" ("almost like dreams") (13) and difficult to sell. Art is valued less than functionality. The narrator, by contrast, has his specialized task, with responsibility for 123 robots and their work cleaning deep-sea pipes, and is needed in the new world of Solaris.

The conflation of human societies with single ecosystems is also found in the visual comparisons drawn on in the poem between the aerial views of the Stavgersand and other cities, and the narrator's pictures of his own body undergoing medical treatment. In the opening section of the poem, the narrator meditates on the view of human settlements from above:

> SOMTIIMS aig find og seer an min screen,
> seer an regio Norwg-West,
> picts takat fra ovfen,
> seer all ljus om natt, spots eftr spots
> so tait og komplex, fra Krisand til Bergn.
> SPOTS af ljus i ein sigd,
> som om all saman hengr. EIN sigd
> klar til ou skera gennom all materie
> og all human life.
> [...]
> JA, er det detta vi er?
> EIN vorld af half moons?
> MEN wat er da oren sol?
> (11)
>
> (SOMETIMES I find and look at on my screen,
> look at the region Norway-West,
> pictures taken from above,
> see all the lights at night, spots after spots
> so tight and complex, from Kristiansand to Bergen.
> SPOTS of light in a sickle,
> as if everything fits together. ONE sickle
> able to cut through all material
> and all human life.
> [...]
> YES, is that what we are?
> A world of half moons?
> BUT then what is our sun?)

This has a parallel a few pages later when the narrator recalls the images of his stricken body:

LIKSOM aig hugsr photon-
picts af min eigen kropp
sist an medic 14.6.
VAR staerk sjuk da, tarmar infectat!
"PATOLOGRISSEN," dei seis set var.
HUGSR min eigen botten-torso, explic
ekstrem klar. HUGSR miner
tarmar og hofts spesi. TARMAR som strakk seg
so open og glatt
gennom meg. OG miner hoftsboon
gule, photonisa. OGSO dei liknat
sigdar, aig tenk naw. HALF moons
an kvar side af miner tarmar.
HAF sidan sett mang gang
an detta picts. FORVIRRANDE,
aig syns. AT detta intern
exist i meg, som halft mitt,
halft noko annat. SOM om dei lefr
siner heilt eigne lifes I meg?

(16)

(AND I remember photon-
pictures of my own body
recently in medical 14.6.
WAS really sick then, intestines infected!
"PATHOLO-LEAK," they told me I had.
REMEMBER my own lower torso, explicit
extremely clear. REMEMBER my
intestines and hips especially. INTESTINES which stretched
so open and smooth
inside me. AND my hip-bones
yellow, photonised. ALSO they were like
sickles, I think now. HALF moons
on each side of my intestines.
HAVE looked many times since then
at those pictures. CONFUSING,
I think. THAT those internals
exist in me, are half mine,
half something else. AS if they live
their whole own life inside me?)

Both the interlinked cities that make up human life on planet Earth and the connected components of a single human body are complex systems of interaction. By means of the parallels between these two passages, human life is likened indirectly to parasites that have colonized a host, or to small communities within a much bigger ecosystem that can be comprehended also in its totality. The longstanding humanistic ideal in Western culture, in which humans are independent of the material world around them, is negated here by the comparisons both to systems much bigger (the solar system) and much smaller (insects and bacteria) than those of human societies.

These two passages also introduce the idea of the moons, a recurring motif throughout the poem; moons are in orbit around a bigger planet, kept in place by the force it exerts on them. The sun is implied, at least lexically, by the name of the new colony—Solaris. Gravity is the system that keeps these moons orbiting around their sun:

WAT ne fersvinna kan,
er gravitationen. GRAVITATIONEN
existen ovfr all distans fra mill
sol-agen away.
                                        og strikt back til uss.
LITL so litl, veik so veik;
men ne zero. GRAVITATIONEN
er det minsta wi kan stola an.
DEN er den onli total
kommunicationen i univrs.
            DEN er arbeiden ogso I uss.
ALL haf wi
                        gravitation-
kroppar, tick tock
i oren celln og knokl.
ALL wi arbeiden, all wi spiik: gravitationen.
VEN wi love og hatr: gravitationen.
(21)

(WHAT can never disappear,
is gravity. GRAVITY
exists over all distance from millions of
sun-years away.
                                        and comes strictly back again.
LITTLE so little, weak so weak;
but never zero. GRAVITY
is the least we can depend upon.
IT is the only total
communication in the universe.
            IT is at work also within us.
ALL we have
                        gravity-
bodies, tick tock
in our cells and bones.

ALL we work, all we say: gravity.
WHEN we love or hate: gravity.)

The ultimate life force is gravity, and this is what keeps Rimbereid's systems working, his humans permanently attracted to the sun they orbit, even if they cannot understand the sun in its entirety. Like the sea in Lem's novel *Solaris*, the role of gravity is a key force in its movements and formations, but the true meaning of those movements is beyond the comprehension of the scientists who study it. The role of gravity, and of the sun, suggests that despite the humans' efforts to understand and control their environment, represented by the engineering of their surface world and their plans to inhabit Solaris, they are in fact parts of an eco-technological superorganism driven by primal or planetary forces rather than by the power of their intellect.

Petro-fiction on the North Sea: Anthropocene entanglements

"Solaris korrigert" is a meditation on the enmeshment of humans and oil technology. Technology in the North Sea lowlands has historically been seen as a project for human engineering and control, even now, with the growing recognition of the collision between human and non-human forces complicit in the Anthropocene. In "Solaris korrigert," Rimbereid plays on the idea of a technology that has become intrinsic both to individual human bodies and to human society. Besides the corralling of human life into eco-technological "systms" characterized by their connectivity and hierarchy, the narrator's physical self is also colonized by technology. He talks about his "nanofingren" ("nano-fingers") (9) and the experience of being injected with "electric pico-molekylen":

VAR an medic i 14.6 for ein veek sidan.
EIN spruta i armr aig fekk, vid electric pico-
molekylen all must hafa
som til *Seifa botten* fara ska.
(38)

(WAS at medical in 14.6 a week ago.
ONE injection in my arm I got, with the electric picto-
molecules all must have
who are to go to the *Safe Seabed*.)

Humans have become the cyborgs of Donna Haraway's famous essay—beings no longer separable from the realm of technology. Haraway's cyborgs are a wish-projection of modern (late twentieth-century) thought, a new fusion of human and machine able to overcome stultifying binary categories. But both post-humanist and materialist thinkers would argue that it has ever been thus; that humans have always been entangled with their own material environment, rather than having dominion over it. Post-humanists consider that the fusion of humans with their technology is nothing fundamentally new, merely an extension of an older entanglement with their environments:

> Notwithstanding the evident uniqueness of homo sapiens, post-humanists remind us that we are not only biophysical entities made of flesh and blood (like other living creatures), but remain reliant on and daily affected by a plethora of non-human substances and actors (e.g. metals, domestic pets, and water). Post-humanists reject the historical argument, famously made by Francis Fukuyama in *Our Posthuman Future* (2002), that we are now becoming post-human by virtue of new technologies such as genetic modification. Their point is that we have always been (and remain) post-human: we are constitutively "impure" and worldly.
>
> (Castree *et al.* 2013, 141)

Neo-materialists hold that humanity has never been the architect of its material world, and that the Enlightenment ideal of progress and dominion was a fallacy. In his essay "Against the Anthropocene: A Neo-Materialist Perspective," Tim LeCain argues that fossil fuel use has given humans a delusion of their own power:

> Humans have obviously become quite powerful—powerful enough to populate nearly every habitable niche on the planet and alter its fundamental biogeochemical cycles. But rather than crediting humans alone, neo-materialism suggests that they accomplished these things only at the price of throwing their lot in with a lot of other things, like coal and oil, whose powers they only vaguely understood and certainly did not really control. Likewise, once these partnerships were made, these powerful things began to shape humans and their cultures in all sorts of unexpected ways, many of them not necessarily for the better.
>
> (2015, 4)

This partnership with oil is the basis for the imagined future of both Stavgersand and Solaris and of the backstory of the breeding oil rigs in Miéville's "Covehithe"—a new kind of humanity that has evolved and adapted with the technology it once thought it controlled. In his essay, LeCain argues against what he considers the misguided concept of "The Good Anthropocene," in which twenty-first century humans might be able to harness their understanding of the environment and technology to come up with technical fixes to the crises of global warming, rising sea levels, and overpopulation. In LeCain's view, the idea that humans are separate from the material world is false: "both optimistic and pessimistic views of the Anthropocene often share the conventional modernist belief that powerful humans and their cultures are distinct from the natural material world" (2015, 8). Recognizing the entanglement of humans and technology is a crucial part of understanding the modern dilemma, and this means rejecting the progressive narrative of the European Enlightenment and replacing it with narratives that make clear the diversity of human-material co-evolution.

The revenant oil rigs in Miévielle's petro-story are a manifestation of the partnership between human and material agencies. Technological creations have come to life through non-human agencies; the process of their merging exerts a powerful fascination: "the strange prurience that endlessly turned on monoliths rutting miles down. An inhuman pornography of great slams and grinding, horrified whales veering from where one rig mounted another, warmed by hydrothermal vents" (350). The moment at which these multiple agencies collide and entwine remains invisible, unimaginable, and the hybrid offspring differ from each other, each a unique figuration of the human-techno-biological nexus, "seven of them, all different designs: buildings, supports, struts, derricks. They waited, swaying like new calves, still wet from their tarry sacs, swinging umbrella-sized cranes" (349).

The fusion of material agencies is likewise woven through Rimbereid's petro-poem "Solaris korrigert." The narrator is preoccupied with his own materiality, his fusion with new technology, and the materiality of the world around him. Oil and oil-related technology are central to the landscape of the North Sea, with the empty oil well providing the *Seifa botten* of Solaris; it is presumably also the basis of *grease*;

> GREASE er wat aig haf
> af mine hendr denna morgning.
> GULSVART glinsande grease
> fra miner robot-
> arbeidern, fra deirs beltfoot,
> arms og linkgreip. AIG love
> miner rektangl-robots (1200 × 400 × 350 mm)!
> OLDA og simpl-mekanical som dei er.
> DEI er ne onli part af modern modell-
> novledg. DEI er ogso part af old sea,
> af old pipes og old grease.
> DEI arbeiden
> I unbegrensa links, miner robots,
> so om ein af dei sku i error gaa,
> ein odder kommen, erstatt. DEI er lik maur,
> onli konstructet.
> ALL er linka til kverodder.
> OG dei til meg, lik aig til kloakk-
> pipes undr bridge via dei,
> og aig via dei og kloakk-pipes
> til odder humans.
> MEN vidout grease
> miner robots vul for lengst haf stifna …

GREASE er links total!
FOR greasen glinsen og glossen
I all flatar
gluen og gliden
i all links og vehickel
og greasen gefr all veksande skinn og life.
DU human, so greasen
ven du born!

(24)

(GREASE is what I have
on my hands this morning.
Yellow-black glistening grease
from my robot-
workers, from their beltfoot,
arms and connector grips. I love
my rectangle-robots (1200 × 400 × 350 mm)!
OLD and simple-mechanical though they are.
THEY are not only part of modern model-
knowledge. THEY are also part of old sea,
of old pipes and old grease.
THEY work
in unlimited connectivity, my robots,
so that if one of them should go wrong,
another comes, replaces it. THEY are like ants,
only constructed.
ALL are linked to one another.
AND they to me, like I am to the sewage-
pipes under bridge via them,
and I via them and the sewage-pipes
to other humans.
BUT without grease
my robots would long have got stiff…
GREASE is total connectivity!
FOR grease glistens and glitters,
in all surfaces
glues and glides
in all links and vehicles
and grease gives all growing skin and life.
YOU human, so you were greased
when you were born!)

The emphasis on connectivity—grease as an agent of connectivity, but also the robots' physical connections to historic materials ("part af old sea, af old pipes og old grease")—underlines the entanglement of natural and industrial products with the human. Audunn Lindholm points out the significance of the final ode to grease in this section of the poem:

> I tillegg … er denne strofen en besyngelse av smøreoljen, *grease*.… Ja, på et tidspunkt erklære sågar diktjeget, med en vri på det bibelske ordspråket "Kom i hu menneske, av støv er du kommet, og til støv skal du vende tilbake": "DU human,/so greasen/ven du born!"
>
> (In addition … this verse is a paean to lubricating oil, grease.… Indeed, at one point, the poem's narrator even declares, with a play on the biblical saying "for dust thou art, and unto dust shalt thou return": "YOU human,/ so you were greased/when you were born!")
>
> (Lindholm 2008, 260)

Grease, the lubricator of human and technical relations, is what allows the futuristic society to function.

Incidentally, this biblical paraphrase creates an interesting echo with Donna Haraway's cyborgs, and their sense of their (new) materiality:

> The cyborg does not dream of community on the model of the organic family, this time without the oedipal project. The cyborg would not recognize the Garden of Eden; it is not made of mud and cannot dream of returning to dust.
>
> (Haraway 1990, 192)

Rimbereid's post-humans are cyborgs, and in the entangled materialities of the Anthropocene, the biblical dust has been replaced by the oil that has been so central to the industrial human project.

Future energies

Both of the petro-texts in this chapter centre on the dynamism of the joint agencies of humans and other materials implicated in petroleum technology. Miéville's rigs are reproducing and growing and moving under the sea. In Rimbereid's poem, momentum is provided by the move towards a new kind of energy, a new frontier beneath the North Sea. The narrator is to be sent to the new "deep-sea-world" of Solaris to pipe up geothermal energy to the surface. The movement towards this future is presented as something with little alternative, despite the narrator's ambivalence about it:

> OGSO aig der ned ska snart.
> AIG vid min mekanical
> novledg og miner robots
> for ou arbeida I NEW-DEPT-SEA-VORLD.
> I vorld undr slam, seagrass
> og fish.

ARBEIDA vid geothermal energi
aig ska, piping den up … EIN slik seifa
vorld er kann henda
oren sista chans, dei seis.
OM wi ska tenka intelligenten
og seifa uss self, som human existensen,
wi haf ou profa wat wi profa kan.
OG difor gaa undr?
GAA undr, gaa ned, gaa i jord
vid all wat we haf af novledg!
DER ned, der ska all for human vera, dei seis.
JA, wi kan mastr detta, dei seis.
OG must difor profa detta,
om wi ne ska enda her up
undr danger himml
som historic skuggar.

(30)

(SO I will have to go down there soon.
ME with my mechanical
knowledge and my robots
to work in the NEW-DEEP-SEA-WORLD.
In the world under mud, seaweed
and fish.
WORK with geothermal energy
I shall, piping it up … SUCH a safe
world may well be
our last chance, they say.
IF we want to think intelligently
and save ourselves, as human existence,
we have to try what we can try.
AND therefore go under?
GO under, go down, go into the earth
with all the knowledge that we have!
DOWN there, we will have every kind of human, they say.
YES, we can master this, they say.
AND we must therefore try it
if we don't want to end up here
under the danger sky
as historic shadows.)

The repeated reliance on expert opinions ("dei seis," "they say," reiterated three times in this passage alone), the repetition in the poem of the word "safe," and the emphasis on knowledge and progress are reminiscent of the Industrial Revolution and the drive to progress that brought about the fossil fuel economy.

The optimism on the part of the presumed experts is put in context by our knowledge of our own time, coupled with the ruined ecosystems that the narrator describes elsewhere in the poem. We realize that elephants are extinct in the wild in the imagined future of 2480, as are all big animals, despite the efforts in our own time to preserve their habitats:

> ELEFANTEN er forts biggast,
> seifat i zoo. MEN deirs kroppar
> impossibl som fri.
> EXPERIMENTA Afriq feilat
> EXPERIMENTA Canada feilat.
> (23)

> (THE ELEPHANT is still the biggest,
> protected in the zoo. BUT their bodies
> are impossible in the wild.
> THE EXPERIMENT Afriq failed
> THE EXPERIMENT Canada failed.)

The human drive to find new solutions, new sources of energy, is constantly creating new hybrid materialities, moving onwards in a perpetual cycle of losses and gains. Fredrik Albritton Jonsson has argued for greater attention to the history of energy use in the Global North as a way of understanding the predicaments that the Anthropocene forces us to confront (2015, 60–3). Dipesh Chakrabarty writes that "the mansion of modern freedoms stands on an ever-expanding base of fossil-fuel use" (2009, 209), but looking into the future with Rimbereid, we might want to write this as "the mansion of modern freedoms is constantly being rebuilt on an ever-changing base of energy use." The "petro-spectral" creations of Rimbereid and Miéville are articulations of the changing imagination of how human and material energies interact. Petro-fiction underlines the tension between oil prosperity and planetary fragility, and highlights the human fears that accompany the narrative of technological development and environmental change.

The human narrative of development has been told as a journey towards the light. As we have already discussed elsewhere in this book, the Enlightenment takes its name from the "absolute metaphor" of the light of truth, a literal revelation in human history (see Chapter 2). In "Solaris korrigert," when the narrator is injected with the "electric pico-molekylen," compulsory for all those who are scheduled to go to Solaris, he is literally "enlightened":

> AIG out i streeten gekk.
> SOL skein so staerk, ein ljus total
> gennom disen.
> ALL kvitt!

SOM aig self ein ljus var blitt
og onli ljus.
(38)

(I went out into the street.

SUN was shining so bright, total light
through it.
EVERYTHING white!
AS if I myself had been turned into light
and only light.)

The narrator has to sit down on a bench for a while, overcome by this experience; he says he has to "lukka miner augr/lukka dei, da plutsl so trang til noko annat vera,/vera som ein moon vidout ljus, or ein skugga/vera, ein skugga vidout ansikt" (close my eyes/close them, as I suddenly felt such an urge to become something different/become like a moon without light, or a shadow/a shadow without a face) (39). Accepting the light of progress, of truth, seems to require some kind of sacrifice. This is clarified a few lines later, when the narrator is sent for further safety checks prior to his departure for Solaris, and is told that he has a small defect in his brain, a surfeit of imagination, "ein noko for staerk production/af eigne picts" (a slight over-production/of my own images) (41).

Now, as the narrator arrives on the elevator platform in the middle of the North Sea, from where his descent to Solaris is to begin, he notes "SEAEN nearli red, skummende red,/syns den vera. AF ljus/or af poison?" (THE SEA nearly red, foaming red/it seems to be. FROM light/or from poison?) (43). The proximity of light and poison suggests the possibility of a light of truth open to corruption, a false light. As already noted in Chapter 2, Markus Wilczek has suggested that light as "absolute metaphor" may have become toxic in the Anthropocene, promising not enlightenment, but irradiation. Moreover, when the narrator is inside the elevator in the very last passage of the poem, he hears a final warning from the man performing the safety controls,

"HUGS at ljus staerkt so staerkt exist an *Seifa botten*!"
er sista aig hoyr i moment door glir igen
og eig tar til ou fara ned.
(44)

("REMEMBER that light strong so strong exists on the *Safe Seabed*!"
is the last thing I hear as the door slides shut
and I begin to descend.)

The poem ends with the narrator's arrival in Solaris; the elevator stops with a click and the last line is "kvar vegg umkring meg fell" ("the walls surrounding me fell away") (45), the emptiness of the remaining page perhaps suggesting the blinding light into which the narrator has been dispatched. The light that beckons with truth and progress may in the end only signal humanity's next entanglement with the fate of the Earth.

Notes

1 The Tourist Trail. [online] Available at www.thetouristtrail.org/guides/suffolk-guides/covehithe/. (Last accessed 16 August 2018.)

2 "Rimbereid har trukket en språklig sirkel rundt Ekofisk, denne første plattformbyen midt i Nordsjøen, og skapt en hybrid syntese ab stavangerdialekt, lavlandsskotsk, engelsk, nederlansk, og dansk, med oldnorske former blandet inn."

3 Refsum (2010, 89) gives the language used by Rimbereid as being a "fremtidsspråk bestående av elementer fra norsk, engelsk, tysk, nederlandsk og norrønt" (a future tongue consisting of elements from Norwegian, English, German, Dutch, and Old Norse) drawn from contemporary developments in Stavanger dialect as a result of the oil industry.

References

Agnew, John A. 2011. "Space and Place." In *The Sage Handbook of Geographical Knowledge*, edited by John A. Agnew and David N. Livingstone, 316–30. London: Sage.

Andersen, Per Thomas. 2018. *Rimbereids Lyrikk: Hendelser og poetiske øyeblikk.* Oslo: Fagbokforlaget.

Augé, Marc. 1995. *Non-Places: Introduction to an Anthropology of Supermodernity*, translated by John Howe. London: Verson.

Castree, Noel, Rob Kitchin, and Alisdair Rogers. 2013. *A Dictionary of Human Geography*. Oxford: Oxford University Press.

Chakrabarty, Dipesh. 2009. "The Climate of History: Four Theses." *Critical Enquiry* 35, no. 2 (Winter): 197–222.

Ghosh, Amitav. 2016. *The Great Derangement: Climate Change and the Unthinkable.* Chicago, IL: The University of Chicago Press.

Haraway, Donna. 1990. "A Manifesto for Cyborgs: Science, Technology, and Socialist Feminism in the 1980s." In *Feminism/Postmodernism*, edited by Linda J. Nicholson, 190–233. New York: Routledge.

Haraway, Donna. 2016. *Staying with the Trouble: Making Kin in the Chthulucene: Experimental Futures.* Durham, NC: Duke University Press.

Hennig, Reinhard. 2014. *Umwelt-engagierte Literatur aus Island und Norwegen: Ein interdisziplinärer Beitrag zu den environmental humanities*. Frankfurt am Main: Peter Lang.

Ingwersen, Moritz. 2017. "'Geological Insurrections:' The Weird Return of Rust and Dust in China Miéville's 'Covehithe' and 'The Dusty Hat.'" Paper given at ASLE Conference 2017, unpublished.

Jonsson, Fredrik Albritton. 2015. "Anthropocene Blues: Abundance, Energy, Limits." In *The Imagination of Limits: Exploring Scarcity and Abundance*, edited by Frederike Felcht and Katie Ritson. *RCC Perspectives* 2, 55–63.

Le Guin, Ursula K. 2015. "Three Moments of an Explosion by China Miéville—Masterfully Horrific SF." *Guardian*, 29 July 2015. www.theguardian.com/books/2015/jul/29/three-moments-explosion-china-mieville-masterfully-horrific-sf. Last modified 22 February 2018.

LeCain, Timothy J. 2015. "Against the Anthropocene: A Neo-Materialist Perspective." *International Journal for History, Culture and Modernity* 3 (1): 1–28.
Lem, Stanisław. 1987. *Solaris*. New York: Harvest.
Lindberg, Elin. 2007. "Ein place millon seagrass og sol: Stedsfornemmelser I Øyvind Rimbereids dikt 'Solaris korrigert.'" Master's thesis in literary studies, University of Oslo.
Lindholm, Audunn. 2008. "Audunn Lindholm om solaris korrigert." In *Norsk Litteræskanon*, edited by Stig Sæterbakken and Janike Kampevold Larsen, 355–68. Oslo: Cappelen Damm.
Madsen, Claus K. 2015. "To aspekter af helhed og to typer langdigte: En afprøvning af Jakobsons poesiformalisering på Høecks *Ulrike Marie Meinhof* og Rimbereids *Solaris korrigert*." *Edda* 102 (4): 296–310.
Miéville, China. 2015. *Three Moments of an Explosion: Stories*. New York: Del Rey.
Mønster, Louise. 2009. "At finde sted: En introduktion til stedbegrebet og dets litterære potentiale." *Edda* 96 (4): 357–71.
Murphy, Patrick D. 2001. "The Non-Alibi of Alien Scapes: SF and Ecocriticism." In *Beyond Nature Writing: Expanding the Boundaries of Ecocriticism*, edited by Karla Armbruster and Kathleen R. Wallace, 263–78. Charlottesville: University of Virginia Press.
Pedersen, Bernt Erik. 2015. "Det store diktet om en framtid uten oljen." *Dagesavisen*, 16 October 2015. www.dagsavisen.no/kultur/det-store-diktet-om-en-framtid-uten-oljen-1.420611. Last updated 19 October 2015.
Refsum, Christian. 2010. "Flerspråklighet i nyere skandinavisk litteratur: Jonas Hassen Khemiri og Øyvind Rimbereid." *Edda* 97 (1): 81–95.
Rimbereid, Øyvind. 2005. "Solaris korrigert." In *Seine topografier/Trådreiser/Solaris korrigert*, 9–45. Trondheim: Gyldendal.
Rimbereid, Øyvind. 2006. "Om det topografiske diktet: Eller i stedet for en poetikk." In *Hvorfor ensomt leve?* 79–244. Trondheim: Gyldendal.
Selås, Jon. 2015. "Teateranmeldelse: 'Solaris korrigert.'" *VG*, 16 October 2015. www.vg.no/rampelys/i/LK2Pq/teateranmeldelse-solaris-korrigert. Last updated 16 October 2015.
Steinkjer, Mode. 2015. "En fabelaktig visjon." *Dagsavisen*, 19 October 2015. www.dagsavisen.no/kultur/en-fabelaktig-visjon-1.421661. Last updated 19 October 2015.
Sullivan, Heather I. 2017. "Material Ecocriticism and the Petro-Text." In *The Routledge Companion to the Environmental Humanities*, edited by Ursula Heise, Jon Christensen, and Michelle Niemann, 414–23. Oxford: Routledge.
Wilczek, Markus. 2014. "When Our Future Is at Stake: A Poetics of Radioactivity." Talk given at the Rachel Carson Center Lunchtime Colloquium on 31 June 2014. www.youtube.com/watch?v=GElzwXx2xsE (last accessed 24 March 2016).
Wilson, Edward O. 1991. "Ants." *Bulletin of the American Academy of Arts and Sciences* 45 (3 December): 13–23.

Conclusion

The literary imagination in the environmental humanities

In the introduction to this book, I argued for the significance of the imagination of the North Sea Lowlands as a bellwether for understanding cultural perceptions of climate change and the planetary crisis of the Anthropocene. The use of topography—in this case, a transnational tidal drainage basin—to delineate a corpus of texts for comparative literary analysis is a novel one, and may inspire new comparative literary studies of landscape features and ecological processes. As a way of exploring the literary imagination of a physical landscape, it opens up new axes for comparison, transcending linguistic, national, and genre distinctions in order to foreground the discussion of material worlds and environmental change.

This book draws on environmental history and ecocriticism in its attempt to ground comparative methodologies within the Environmental Humanities. The new field of Environmental Humanities calls for inter- and transdisciplinary approaches: a holistic humanities to try to grasp the vast and interlinked ecological, biophysical, and geological processes of our common planet, which we both are subject to and are shaping in new ways. Scholars in this field must draw on a broad range of materials and methods and move across different areas of enquiry to investigate research questions that are pertinent to a state of environmental crisis. In turn, then, this book seeks to speak *back* to the disciplines it has roamed across, and submit its insights to the field of Environmental Humanities. In particular, it offers environmental historians new ways to use comparative literary analysis in their studies of the past.

Poetry, literary prose, and novels are part of the system of human culture by which texts reflect and co-constitute human ideas about the natural world. Part of a broad and open-ended literary discourse, in practice they rarely cross over as sources for historians. Within literary studies, the contexts in which literary texts are examined are often ones dictated by conventions from national literary traditions. The axes of comparison favoured by comparativists are often, as discussed in the introduction to this book, literary-reflexive ones; comparisons are made between different examples of a particular genre, or stylistic features such as narrative voice, usually across cultural, temporal, or linguistic divides.

My approach in this book has been informed by the way that historians use their sources, "mining" a range of very different types of text for the insights they can give to the perception of a physical and cultural space. The work with sources thus combines the techniques of literary close reading with a broader approach to literary texts that foregrounds particular themes and discourses about the natural world; genre, language, and style still inform my close readings, but are not treated as defining categories. My reading across five languages and a range of genres and forms to identify common ideas and discourses within them adds to historical and geographical understandings of North Sea places, in particular the ways in which they are imagined as landscapes and seascapes in the context of global environmental change.

Ideas about landscape are subjective and shifting, contingent on land use practices, on economics, on ideas of conservation and biodiversity, and on the emotions evoked by a particular place. Environmental historians know this. They track the ways that landscapes are accorded value over time, as recorded in sources as diverse as settlement patterns, land deeds, conservation laws, and flood markers. But the ways in which landscapes and ecologies make their way into our literary texts, the ways in which they are imagined and constituted in literary discourse; these too offer insights for environmental history. Environmental historians are moreover sensitive to the agency of nature, finding between the lines of their written sources evidence of ecological processes often understood only with hindsight. The landscape imaginary articulated in literary form requires the same kind of reading between the lines: in close readings of literary texts, we can find traces of an understanding of more-than-human agencies, and discursive reflection on the relationship between humans and their environment.

The North Sea lowlands are real places, silty and shifting, but they are also places of the human imagination. I have referred to them in this book mostly as landscapes, in order to emphasize their co-construction both as physical places and as human imaginaries. The word *landscape* is an aesthetic term, describing a place that is referenced by the human eye that sees it. Landscape painting, along with other media and text, has shaped our ideas about the places we value in immeasurable ways. In Western Europe, for example, the role that the Romantics played in influencing our ideas of what constitutes landscape was particularly significant. The Romantic sublime mode created iconic images, from Constable's and Wordsworth's portrayals of England's pastoral idylls to Caspar David Friedrich's monumental waterfalls and mysterious forests, and these still serve as benchmarks by which beauty and worth are measured; influencing how we demarcate the environments we live in from the landscapes that we see. Landscape only comes into being once there is a human eye to see it, and once we see a physical place, we enter into a relationship with it that is mediated by our own particular aesthetic and cultural ideas about landscape.

The environmental imagination is concerned with physical processes and events, as well as places; landscapes are temporal—they have pasts and futures. When a terrible storm flood strikes, the memory is preserved not just in flood markers and raised dykes, but in the (re)imagination of past and future floods. These historical imaginaries can be aestheticized and instrumentalized politically in the service of nation-building, or engineering projects, or they may be suppressed and ignored when they run counter to narratives of progress. In an age where the threat of climate change is keenly felt, the imagination of future floods is an important tool in understanding the real places we inhabit. Literary texts are a constitutive part of such landscape imaginaries.

The value of the North Sea Lowlands as both a natural and cultural site is perceived and configured in dialogue with multiple imagined lowlands. Migrating birds gradually disappear from the wetlands and reappear in literary prose; low-lying ground on which houses and schools stand are, in imaginary floods, inundated and lost to the sea. The narratives and counter-narratives that emerge in the interaction between landscape and literary text can inform the work of historians, and of scholars in the Environmental Humanities more generally who want to look deeply into the way the North Sea coasts are known and understood. Literary interventions play a part in ascribing value to our landscapes, and they both reflect and constitute human ideas about the lowlands over time. Literature foregrounds the interaction between humans and nature, texturing the landscape with a range of human motivations and emotions. In the following section, I will explore three insights into the imagination of the North Sea lowlands that come out of the close readings in this book, and show how the inclusion of literary scholarship can enrich and complicate a historical understanding of these places.

Amphibious cultures

Historians such as Petra van Dam and Greg Bankoff (van Dam 2016; Bankoff 2013) have made the argument that the cultures of the North Sea lowlands share particular ways of living with water. Laws regarding dyke construction and maintenance, the practices of flood memorialization, and an understanding of the tides are part of what van Dam and others describe as an "amphibious" way of life. Bankoff uses the idea of disaster culture as an organizing principle to describe communities that have developed a shared understanding of flooding hazards, and conceptualizes a specific North Sea risk culture that is common to the coastal lowlands of the North Sea basin.

Literary work by leading figures in their respective national contexts—Hans Christian Andersen in Denmark, Arthur van Schendel in the Netherlands, and Theodor Storm in Germany, for example—show how powerful the historical imagination of such amphibious societies can be. The ability to recover from past storm floods and protect against future ones are critical features of a society attuned to its environment. Such events can serve not just as literal reminders of similar episodes that have happened and might happen again, but metaphorically, serving as creation myths for social cohesion, or as external threats to the status quo.

In the novels *The Two Baronesses* and *The Waterman*, we see how the interrelationship between human society and the natural world can be mapped onto the struggle of the nation-state in times of conflict and consolidation. In their writing, both authors were responding to very real environmental conditions. The Netherlands suffered repeated flooding during the eighteenth century, when its resilience had been lowered by war and occupation, and Denmark in 1848 was only just recovering from the effects of deforestation and sand storms. These real events informed literary works, which then themselves shaped national attitudes to human resilience and natural agency in those places. Denmark was inscribed into culture as a state of mind, pure and free, that was elevated from shifting political borders or, quite literally, shifting sediments. The Netherlands could assert itself as a nation only when its downtrodden society had learnt to organize against the threats of its climate and watery situation. The landscapes we find in literature are never devoid of context; literary works are written and read in shifting tapestries of historical, national, political, and ecological awareness and conviction.

This conflation of specific regional conditions with national character has had a complex afterlife in the Netherlands. Comparing *The Two Baronesses* with subsequent texts that reflect on amphibious societies and their ability to withstand disaster suggest a long and ongoing process of dialogue between the real and imagined North Sea lowlands. Jan de Hartog's novel *The Little Ark*, like other scattered literary responses to the 1953 flood disaster, rejects any central narrative of strength and resilience in favour of one that is, in sum, bewildered and uncomprehending. Later still, Jim Shepard's short story, "The Netherlands Lives with Water," satirizes modern-day Dutch political rhetoric, which proclaims an amphibious society that is conscious of the risks to which it is exposed, while climate change raises the stakes. Most of the residents of Shepard's future Rotterdam, prosperous and dry, live in wilful denial of the rising waters from which even the most talented and well-funded hydro-engineers are unable to protect them. These texts provide a critical counter-narrative to the political forces that constructed the Delta Works and continue today to insist that the Netherlands is able to adapt to climate change.

In the case of Theodor Storm's *Der Schimmelreiter*, the amphibious culture inhabited by his creation Hauke Haien is a site of conflict between Enlightenment idea of human reason and the animism and superstitious Christianity that it had set out to supersede. The novel shows again the power of the amphibious society—here standing in for any human culture that has adapted to a particular environment—against the modernizing ideal of progress and the subjugation of the natural world. In his novel, Storm captures the complexity of the transition from an amphibious culture to one based on minimized risks and increasing prosperity: it is a veritable parable of unintended consequences, which does not preclude its later misinterpretation in the service of political agendas. Storm's Frisian coastline is a battleground for opposing forces, the drive of the new and the pull of the old, and the metaphorical power of land being reclaimed from the hungry sea has persisted, most notably with the incorporation of Frisian land reclamation into the *Blut und Boden* rhetoric of the National Socialists with the creation of Adolf-Hitler-Koog, Hermann-Göring-Koog, and Horst-Wessel-Koog (all since renamed) in the 1930s.

Literature exists in a political and historical context, and it is an important tool for the political imagination. But it can also provide a forum, by means of techniques such as ambivalence, satire, and pastiche, for a subversive commentary of political developments. Amphibious societies exist not just in the lowlands, but also in the shared imagination of them. The challenges faced by those who inhabit vulnerable littorals are shaped by the way they see themselves and understand their role in wider society. In our current moment of interlocking environmental crises, a critical understanding of places that pays attention to the way they are portrayed and constructed in literary discourse is a vital part of the Environmental Humanities.

Erosion and loss

The environments of the North Sea lowlands are, today, doubly threatened. Erosion and rising seawater have been eating away at the coastlines for millennia in a process that has often been dramatic and all too visible. But the processes of the Anthropocene have not only accelerated the pace of sea-level rise, but have also added pollution, overfishing, warming temperatures, and biodiversity loss into the mix. How have these twin vulnerabilities, old and new, been experienced and understood?

The North Sea lowlands' longstanding vulnerability in the face of seawater encroachment and land erosion are visible in the landscape. Dykes and flood markers are signs of inundations; the mudflats at low tide can reveal the dykelines of previous centuries, and in places medieval wooden dyke-posts and the grassy remains of sheep pasture waving in the brine. The landscape is a palimpsest, sediment upon sediment, carrying traces of the past that are visible in the present. Striking juxtapositions, such as church spires disappearing beneath the waves and gravestones aslant on the edge of crumbling cliffs, are landscape features that stand synecdochally for the region as a whole.

These traces of former landscapes are also found in literary portrayals of the region, imbued with the symbolism of lost worlds. The earliest literary text explored in this book, Hans Christian Andersen's *The Two Baronesses*, contains an ecologically-powered vanitas motif, when the servant girl Keike moves the bones from the graveyard at the eroding end of the island to a new spot at the other end. Unsurprisingly, similar motifs are an important part of the imagination of the North Sea lowlands. The sunken medieval city of Dunwich is found in both W.G. Sebald's *Rings of Saturn* and Robert Macfarlane's "Silt," and a strikingly similar fictional sunken city is rowed over by the protagonist of Meg Rosoff's *What I Was*.

The idea of the palimpsest is one that comes from literature; it describes the ghosts of former texts that resist total overwriting. And literary portrayals of the North Sea coasts are palimpsests both in their descriptions of landscape and in their own intertextuality. We see this at work most clearly in the genre of new British nature writing explored in Chapter 4 of this book. This form of writing gives us descriptions that are both literary and literal (as well as littoral) palimpsests; these writers read the landscape both materially and textually, referring to material traces of the past they find in the landscape itself and traces of the landscape's past in other texts. They show the way that the imagined landscape is both made up of and mediated not just through its physical existence but also its textual record, which forms the sediments of our human imagination. Human settlement and human history has added layer upon layer to the landscape, meanwhile, erosion is one process that uncovers and wears away, revealing what lies hidden beneath.

The interaction between littoral and literary reveals the way that landscape and text inform and influence one another. The sheer fact of the North Sea lowlands becoming a popular topic for nature writing testifies to a new appreciation of these flat, marginal spaces: writing and reading about a place is a way of inscribing it into the literary imagination, giving it meaning and value. East Anglia is drying out, shrinking, eroding, and it is ever more vulnerable to tidal movements, but its hastened decline is marked by its increasing consolidation and centrality in nature writing and fiction. Land that is reverting to mudflat and seabed is being reclaimed in the imagination. Birds and grasses that have been disappearing from the shores over the past 50 or so years are named and described and mourned in literary form, creating a textual palimpsest of a landscape that will be traceable even after the last bird has flown away. Literature can show the history of place using its own layering of intertextual sediments—rhetorical structures of synecdoche, metaphor, repetition, and lexical variations that pattern change through the human medium of language.

Historians who read the past of the landscape together with the history of the humans who lived in it can find in literary texts evidence for the interaction of the human imagination with the physical place. Moreover, historians who employ their imaginations of past worlds to produce written texts of their own are embroiled in the same interactive processes between text and place. How we read and understand the landscape is mediated by our imagination; of what was there before, and also—now more than ever—what will be there in the future. The sensitive Claus in Kjersti Vik's novel *Mandø* imagines for us a truly ancient palimpsest, discernible in the landscape only to geologists, of the Earth when it was a young blue planet of roiling waters, uninhabited by humans; being able to "read" this palimpsest through the many-layered landscape of our imagination is something that I hope will serve the Environmental Humanities well as we head into the uncertain future of the Anthropocene.

The Anthropocene: Chakrabarty's "Four Theses"

In this book, I have returned again and again to Dipesh Chakrabarty's "The Climate of History: Four Theses" to illuminate my analysis of literary texts in the context of thinking on the Anthropocene. Chakrabarty's texts have spawned long discussions and a great deal of scholarly responses, but his outlining of the paradoxes of the Anthropocene as they struck him in 2009 remains a rallying point for Anthropocene scholars everywhere. Literary texts can be useful when thinking with and responding to the paradoxes Chakrabarty outlines in trying to grasp the abstract dimensions of the Anthropocene. Literature has the potential to make the abstract concrete, to ground questions in places and people. The choice of subject matter—for example the focus around oil and oil technology in the works by China Miéville and Øyvind Rimbereid—is an explicit engagement with the idea of humans as *geological* agents, rather than merely biological ones; the disturbing agency of oil is made alarmingly concrete in Miéville's walking, breeding oil rigs. The futuristic scenarios in these two texts make clear that geological agency and geological intention are not the same thing. But, more than that, literature can both contain and generate great interdiscursivity; not only can literary works bring Chakrabarty's theses to life, but they can bring them into conversation with other discourses and ideas—with Donna Haraway's cyborgs, for example, and with Tim LeCain's Carbocene, with biologists' theories about superorganisms, and with the idea of the human subconscious and suppression of fears that has its roots in the thinking of Sigmund Freud. This interdiscursive richness allows for multiple readings; the Environmental Humanities, as a field concerned with making sense of overlapping and entangled narratives and processes, can make use of this interdiscursivity in productive ways. Placing Chakrabarty's theses in tension with others, in concrete imaginations of places and things, provides scholars of other disciplines with new ways of thinking with them.

Rimbereid fuses the geological with the technological and human in his lyrical aestheticization of "yellow-black glistening grease" on the hands of his narrator, which links him to the robots he commands. In his first thesis, Chakrabarty charges environmental historians with too often framing histories in terms of the interactions between humans and the natural world in a way that reinforces the separation between the two, rather than allowing the slippage that the Anthropocene represents. Doubtless this is not always true; but environmental historians who are open to the realm of the environmental imagination of the places and processes they study—as embedded in literary texts and in other cultural products—would find it more difficult to ignore the ways that human history and planetary geology are entangled with each other. Neomaterialists have argued this, but in literary texts such as Rimbereid's, the abstract idea of entanglement is made manifest—sticky, in this case, a coating of grease (Rimbereid 2005, 24).

Likewise, Chakrabarty's contention that the history of human Enlightenment was powered by the fossil fuel revolution—thereby "severely qualif[ying] human histories of modernity"(Chakrabarty 2009, 207)—is given flesh and blood in literary form. In particular, Theodor Storm's masterpiece *Der Schimmelreiter* can be read as a problematization of the Anthropocene *avant la lettre*. The forward-thinking dyke engineer is not just blind to the environmental consequences of building new dykes that override traditional knowledge of the power of water and waves on his stretch of the North Sea coast; he must also make the bitter discovery that human progress requires a sacrifice, human blood, in a work of narrative circularity that refutes the linear, open-ended concept of progress. The ambivalence of this text, which both undermines and reasserts religion and the narrative of progress in its spiralling structure as well as its subject matter, encapsulates the work of historiography today in seeking to unpick and reinterpret the history of the modern period. *Der Schimmelreiter* also demonstrates how literary texts can operate discursively both in terms of the subject-matter and in the medium of its expression: narrative structure and voice in the novella function as counterpoint to plot and character to question assumptions of linearity and closure.

Literatury texts can also answer Chakrabarty's question of how to "create a conversation" (214) between deeply polarized human populations and the humanity that is a planetary species engaged in the process of destroying its own habitat. W.G. Sebald's work attempts just this: his labyrinthine prose incorporates a multi-scalar view of history, oscillating between deep time and recorded history, between stories of individuals and stories of massed humanity. His use of distortion, digression, and unreliability, both in the form of the narration and in his many illustrations and descriptions of views and vantage points, emphasize the difficulty and contortion involved in reconciling the overlapping scales of human, ecological, and geological history. Sebald's careful threading of the stories of herring and silkworm through his text also question the anthropocentrism of Chakrabarty's theses; the Anthropocene is not just about the paradox of differentiated humanity versus the human species, after all, nor just human history versus geological history, but also the many shades of ecological histories and species histories that are implicated in these. Here, literary interdiscursivity allows for a critical reflection.

Chakrabarty situates the difficulty of understanding ourselves on the level of a species at "the limits of historical understanding" (221); the ability to historicize ourselves as a species might be a step towards being able to think and act as a species to temper the effects of the Anthropocene, but according to Chakrabarty, it is one that is categorically nearly impossible to undertake. What I hope this book shows is that such a feat of the imagination, if we are to attempt it, cannot be achieved by historians alone. The insights and techniques of those who work with the artefacts of our cultural imagination—the inscriptions of our fears, dreams, confusions, grief, and longings—are a critical resource for all scholars of the Environmental Humanities who seek to probe the limits of the anthropocentric understanding of the world.

*

Shifting Sands puts the fragile and engineered landscapes of the North Sea lowlands at the heart of our environmental imagination at a time in which we fear for the future of human life on Earth. It argues for the importance of the literary imagination in understanding the history, present, and future of our landscape, and in trying to think through and with the many scales and tensions of the Anthropocene.

The literary texts collected and studied here reflect on how attempts to tame the sea and win the land have become entangled with natural processes of shifting sands and unpredictable tides, and with the geological processes that underpin them. They weigh the costs of human intervention in the landscape, the perceptions of risk and disaster, the price of progress and prosperity, the relationship between marginal spaces and global processes, the need for new ways of thinking, new memories, and new imaginaries of the future.

The North Sea lowlands constitute a space in which the relationship between humans and the natural world has been and continues to be contested. In our new age of uncertainty, the ways in which we understand these vulnerable landscapes will continue to reflect the state of a human-nature relationship that is as transient and migratory as the shifting sands on which it is built.

References

Bankoff, Greg. 2013. "The 'English Lowlands' and the North Sea Basin System: A History of Shared Risk." *Environment and History* 19 (1): 3–37.

Chakrabarty, Dipesh. 2009. "The Climate of History: Four Theses." *Critical Enquiry* 35, no. 2 (Winter): 197–222.

Rimbereid, Øyvind. 2005. "Solaris korrigert." In *Seine topografier/Trådreiser/Solaris korrigert*, 9–45. Trondheim: Gyldendal.

van Dam, Petra. 2016. "An Amphibious Culture: Coping with Floods in the Netherlands." In *Local Places, Global Processes*, edited by Peter Coates, David Moon, and Paul Warde, 78–93. Oxford: Oxbow Books.

Index

Page numbers in **bold** denote figures.